Flow cytometry is now well established in research laboratories and is gaining increasing use in clinical medicine and pathology. The technique enables multiple simultaneous light-scatter and fluorescence measurements to be made at the individual cell level at very rapid rates and results in very large quantities of data being collected. Data, however, are just a series of numbers which must be converted to information which, in turn, must be shown to have meaning. This is the most important single aspect of flow cytometry, but it has received relatively little attention. One of the frequently voiced advantages of the technology is that it produces "good statistics" because large numbers of cells have been analysed. However, it is not very often that confidence limits are placed on results, and hence the reader has little or no feel for the inherent variability in the information produced.

This book covers very basic number-handling techniques, regression analysis, probability functions, statistical tests and methods of analysing dynamic processes. These are developed for the analysis not only of individual DNA histograms to obtain the proportion of cells in the cell-cycle phases, but also of time courses of DNA histograms to yield cell-cycle kinetic information, overlapping immunofluorescence distributions with confidence limits for the estimated proportions, and enzyme kinetic and membrane transport parameters. A brief introduction to multivariate analysis is also given. A distinction is made between data handling – for example, gating and counting the numbers of cells within that gate – and data analysis itself, which is the means by which information is extracted.

All those who use flow cytometry in their research will find this book an invaluable guide to interpreting the data produced by flow cytometers.

T0275568

Flow cytometry data analysis
Basic concepts and statistics

Flow cytometry data analysis
Basic concepts and statistics

James V. Watson

Clinical Oncology Unit
Medical Research Council
and
Faculty of Clinical Medicine
The Medical School
University of Cambridge

CAMBRIDGE
UNIVERSITY PRESS

CAMBRIDGE UNIVERSITY PRESS
Cambridge, New York, Melbourne, Madrid, Cape Town, Singapore, São Paulo

Cambridge University Press
The Edinburgh Building, Cambridge CB2 2RU, UK

Published in the United States of America by Cambridge University Press, New York

www.cambridge.org
Information on this title: www.cambridge.org/9780521415453

First published 1992
This digitally printed first paperback version 2005

A catalogue record for this publication is available from the British Library

Library of Congress Cataloguing in Publication data
Watson, James V.
Flow cytometry data analysis : basic concepts and statistics /
James V. Watson.
p. cm.
Includes bibliographical references and index.
ISBN 0-521-41545-4
1. Flow cytometry – Statistical methods. 2. Flow cytometry –
Mathematics. 3. Flow cytometry – Data processing. I. Title.
QH324.9.F38W37   1992
574.87′028 – dc20   92-14761
                                                                CIP

ISBN-13 978-0-521-41545-3 hardback
ISBN-10 0-521-41545-4 hardback

ISBN-13 978-0-521-01970-5 paperback
ISBN-10 0-521-01970-2 paperback

Contents

1

Introduction

Flow cytometry is now a well established technique in cell biology and is gaining increasing use in clinical medicine. The major applications to date in the latter have been in DNA histogram analysis to determine "ploidy" (DNA index, Hidderman et al. 1984) and S-phase fractions for prognostic purposes in cancer patients and in immunophenotyping (Parker 1988). However, more recent applications in cancer work include determination of tumour cell production rate using bromodeoxyuridine (Begg et al. 1985) and estimations which relate to therapy resistance including glutathione, drug efflux mechanisms and membrane transport (Watson 1991). The power of the technology relates to its capacity to make very rapid multiple simultaneous measurements of fluorescence and light scatter at the individual cell level and hence to analyse heterogeneity in mixed populations.

The early commercial instruments were somewhat fearsome beasts with vast arrays of knobs, switches, dials, oscilloscopes and wires hanging out all over the place. At best, they tended to be regarded as "user non-friendly" and at worst as "non–user friendly". However, the recent generation of machines have been simplified considerably, with the in-house computer taking over many of the tasks which the operator previously had to perform manually. The undoubted "user-friendliness" of these modern instruments, together with the relative reduction in initial capital outlay, is a considerable advantage as it makes the technology available to many more users. In turn, this makes it possible for relatively untrained persons, who may not be fully aware of potential problems and pitfalls, to stain samples and operate the instruments to produce "numbers". There appears to be a prevalent philsophy amongst the instrument manufacturers to produce bench-top devices that require a minimum of operator interaction, so that all that needs to be done is stain up the cells, shove them in the instrument, and out come the numbers.

I'm sure this philosophy is fine from the manufacturers' standpoint, as this approach helps to sell more machines because you purchase an instrument specifically designed to do a particular task. Under test conditions the instrument will perform very well the particular task for which it was designed. However, there are a number of disadvantages to this philosophy. First, a particular instrument designed specifically for a particular task may not do so well with an apparently similar task using different combinations of fluorochromes for different purposes. Second, the "new" generation of flow cytometry users and operators may not even be aware that such problems could exist. Third, the operator is usually insufficiently aware of deficiencies or potential deficiencies in a particular instrument, as no manufacturer will ever say it's not very good at doing this or that. Finally, many operators are completely at the mercy of the software data-handling package supplied, which may contain deficiencies that the manufacturers do not appreciate.

Flow cytometers produce a vast amount of data, which is one of their many attractions, but this can be a two-edged sword. Data, which are just a series of numbers, must be converted to information. Moreover, the information produced from those numbers not only must have meaning, but also must be shown to have meaning. This is the most important single aspect of flow cytometry, but it has received relatively little attention.

One of the frequently voiced advantages of the technology is that it produces "good statistics" because large numbers of cells have been analysed. However, confidence limits are seldom placed on results, and hence the reader has little or no feel for the inherent variability in the information produced. This variability is important and has three major components. The first is due to the measuring system, and applies not just to flow cytometry but to every measurement system. Manufacturers will tend to downplay or ignore this component. The second component is due to variability in the processes involved in making the measurement possible, and in flow studies this includes variability in fluorescence staining procedures (including the various reagents) as well as the technical competence with which the procedures are carried out. The last, and most important, source of variability is within the biology being studied, and it is from this that we might gain some extra information.

This short monograph was compiled from a series of notes originally intended for users of the custom-built instrument in the MRC Clinical Oncology Unit at Cambridge. All of the procedures described in this book are contained within our computer analysis package, which has been updated continuously over the past decade, and most of the examples are drawn from our data base. The statistics sections are limited to those we have found most useful, but I hope this will provide newcomers to flow cytometry with insight into some of the potential problems to be faced in data handling, data analysis and interpretation. The statistics begin with some very basic concepts including measurement of central tendency, distances between points and hence assessments of distributions, and what these various parameters mean. These basic concepts, of which everyone is well aware, are then developed to show how they can be used to analyse data and help convert them into information. A distinction is made here between data handling – for example, gating and counting the numbers of cells within that gate (a process commonly regarded as data analysis but which, in reality, is data handling) – and data analysis itself, which is the means by which information is extracted. Gating is not covered as a specific topic.

The book is intended for biologists using flow cytometers who know about the basic anatomy and physiology of these instruments. Data analysis obviously implies that mathematical ideas and concepts will have to be considered. However, as the book has been written for biologists, an attempt has been made to make this aspect as simple as possible; if you can add, subtract, multiply, divide and, most importantly, think logically then you should have few problems. If you are also familiar with power functions, logarithms, transcendental functions, differentiation, integration and basic statistics then you probably need not be reading this. This book is not intended for highly experienced users and developers with backgrounds in physics, mathematics or statistics who also have struggled with the various problems considered within these pages.

2

Fundamental concepts

Handling and interpreting numbers is not generally a "strong point" for biologists. This applies particularly when there is a large group of numbers that are more easily handled and understood by using some average value as a summary. Indeed, large groups of numbers must be handled by some sort of summary system, because just supplying the raw data – say, 10,000 fluorescence or light-scatter recordings from a flow cytometric analysis run – would be essentially unintelligible. This chapter was included to re-familiarise the reader with some very basic number handling concepts and to set the scene for converting numbers into information.

2.1 Central tendency

If we make a number of measurements on a population, say the weights, shapes and various dimensions of 1000 females between the ages of 15 and 25, we will end up with a large series of numbers. I have chosen this particular example not because I'm a male chauvinist pig, but because I was recently talking with a designer of female undergarments who had the task of converting those numbers into articles for sale on the shelves of a large retail outlet. The end points of the survey were to make the articles as appealing as possible (that was the first consideration), in the minimum number of different shapes and sizes as possible, as cheaply as possible, and to sell all of them all the time. At the outset is was appreciated that some of these aims were mutually exclusive. For example, the minimum number of shapes and sizes is obviously 1, which is the cheapest manufacturing option. A single standard shape and size can be obtained by summing all the measurements in a particular class of measurement, x, and dividing by the number N of females surveyed. This is the arithmetic mean of the population, $\bar{x}$, which is given by the following formula:

$$\bar{x} = \frac{\sum_{k=1}^{k=N}(x_k)}{N},$$

where the symbol $\sum$ represents the summation of all the k x-values from 1 to N.

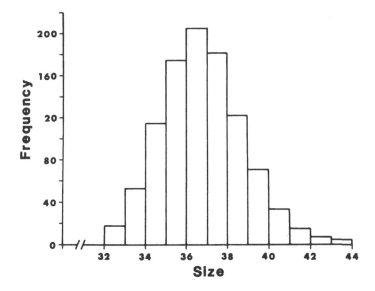

Figure 2.1. Distribution of measurements which exhibits a "skew" to the right.

The problem here is that most of the customers would not be satisfied most of the time, as there are very few who are "average". A further problem, which relates to maturity, is encountered in this age group. There is a relatively greater proportion of smaller sizes than larger, a general phenomenon that applies to the majority of biological observations and measurements. Thus, we cannot represent the whole population by a single set of average measurements as those measurements are distributed; a representation of one set of those data (I'll leave you to guess which) is shown in Figure 2.1, where the distribution is slightly "skewed" to the right. However, we will see later that there are ways of coping with variability of this type.

There are two further methods of expressing central tendency, the mode and the median of the distribution, but neither will help us with the undergarment problem. The mode is the point in the distribution where the frequency is at a maximum, and the median is the point where half the area lies to each side. In symmetrical distributions the mean, mode and median are the same point, but this is not true for skewed distributions; the relationship between these parameters is shown in Figure 2.2.

2.2 Absolute distance between points

Measuring the absolute distance between two points is not as immediately straightforward as it might seem, even though we are dealing with Euclidean as opposed to relativistic distances. Consider two points marked on a sheet of paper. We can place a ruler between these points so that the "zero" mark is adjacent to one point and read off the number on the scale adjacent to the

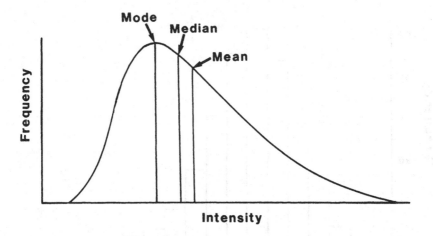

Figure 2.2. Relationships between the mean, median and mode.

second point. Let us suppose the answer is 50.8. If we now take a different ruler we might obtain a distance of 2.0. Which is the correct answer? Obviously, both are correct and you will immediately recognise a calibration difference: the first ruler was metric and the second was imperial. Thus, in order to avoid any possibility of misinterpretation, always state which units you are using. Now, rotate the sheet of paper 180° and repeat the exercise. You will get a very similar, but not an identical, answer. This begs two further questions. Where is ZERO, and why might the results be slightly different? The answer to the first question is that ZERO is exactly where you want it to be. There is no such reality as ZERO; it is a convention just as the whole of the measuring system being discussed is a convention. An answer (but not necessarily *the* answer) to the second question is in Section 2.4.

At first sight the introduction of a discussion about where ZERO might be may seem a little esoteric, unnecessary or even futile, but it is very important in flow cytometry. We do not directly measure whatever commodity we are measuring, we measure light. This strikes the photocathode in the photomultiplier, which then emits electrons in direct proportion to the number of photons striking the cathode. This is the point at which the light flux is transduced (changed) into the electronic signal. The electron flux emitted from the photocathode is then amplified through a dynode chain within the photomultiplier to give a current, thence to produce a voltage which is amplified. Modern electronics are capable of amplifying a signal by many orders of magnitude depending on the various settings you happen to choose, and instruments can have a total dynamic range, in terms of number of molecules measurable, of from 10^2-10^{16} per cell. This range of fourteen orders of magnitude must be viewed through a "window" which, even with log-amplifiers, is still only 10^4 wide. Hence, ZERO on the particular viewing window you are using is an arbitrary point dictated by the assay you are performing, which is somewhere within the 10^2 to 10^{16} total possible dynamic range available to you.

How then can we measure the absolute distance between two points within this arbitrary measurement system, particularly as we do not know where ZERO happens to be located? The square of a positive number is positive and the square of a negative number is also positive. Thus, we obtain the absolute distance between two points by subtracting one from the other (it doesn't matter which from which), squaring the result then taking the square root. This gets around the problem of where ZERO might happen to find itself; it could be on the left or the right or up or down, or in or out, or anywhere, it is no longer of any consequence. This relationship, which is extremely simple and of fundamental importance, is as follows:

$$D_a = \sqrt{(x_1 - x_2)^2},$$

where D_a is the absolute distance between points and where x_1 and x_2 are the two measurements in a 1-dimensional data space.

Let us now suppose we are working in a 2-dimensional data space of (x, y)-coordinates. The absolute distance between any two points with respective coordinates (x_1, y_1) and (x_2, y_2) is given by

$$D_a = \sqrt{(x_1 - x_2)^2 + (y_1 - y_2)^2}.$$

This should have a familiar ring to it, as Pythagoras of Samos did it first in 600 B.C. or thereabouts (it's in the reference list!). It should now be intuitively obvious that we can generalize this relationship for any number of sets of different measurements. Moving up to the 3-dimensional coordinates of the (x, y, z) data space, we have

$$D_a = \sqrt{(x_1 - x_2)^2 + (y_1 - y_2)^2 + (z_1 - z_2)^2}.$$

When we go beyond 3-dimensional coordinates we enter Euclidean hyperspace, which is just a fancy term (but it sounds impressive) to describe sets of four or more coordinates. It is not something that is easily visualized, as we live in and are familiar with only 3-dimensional space; however, extension of the Pythagorean relationship to hyperspace is perfectly valid and was used to solve part of the multidimensional undergarment problem.

2.3 Dispersion and its representation

Just as we must have some way of summarising an average, say the arithmetic mean of a set of numbers, in order to make those numbers "handlable", so we must also have some way of measuring and representing dispersion in that set of numbers. Clearly, we cannot give each individual dispersion, so we must represent this by some sort of average dispersion value. Such a measurement of "spread" is required whenever data are distributed, as there must be some way of expressing how far the overall characteristics of the population differ from the average.

2.3.1 Mean deviation

The simplest measure of dispersion is given by the mean deviation d, where

$$d = \Sigma(x - \bar{x})/N.$$

The expression $(x - \bar{x})$ represents the subtraction of the mean $\bar{x}$ from a given point x to give the deviation of this point from the mean, and the symbol Σ represents a summation of the deviations of all the points from the mean. The summation is then divided by N, the number of observations, to obtain the mean deviation. You will recognise that the mean deviation can be zero if the sum of the deviations above the mean is equal to the sum of deviations below the mean. We will see in Sections 4.2.3, 4.4.1 and 5.2 how this can be used in significance testing.

2.3.2 Mean squared deviation, variance

The mean deviation just described gives equal relative weight to all deviations from the mean, whatever their magnitudes. Another measure of spread, one that gives greater weight to the larger deviations, is given by the mean squared deviation or *variance* σ^2:

$$\sigma^2 = \Sigma(x - \bar{x})^2/N.$$

Thus, the variance is defined as the overall average squared distance (deviation) of all the points from the mean of all the points.

There are a number of ways in which variance is defined, depending on the type of problem and distribution being considered. This particular definition of variance relates to observations or measurements where deviations from the mean are likely to be equal on either side of the mean.

2.3.3 Standard deviation

The standard deviation s is obtained from the variance by taking the square root; hence, it is the root-mean-squared or RMS deviation:

$$s = \sqrt{\Sigma(x - \bar{x})^2/N}.$$

If we now take the square root of the top and bottom of this expression separately, we may write this as

$$s = \frac{\sqrt{\Sigma(x - \bar{x})^2}}{\sqrt{N}}.$$

We can now see that the term $\sqrt{\Sigma(x - \bar{x})^2}$ is a Pythagorean type of expression, where

$$\sqrt{\Sigma(x-\bar{x})^2} = \sqrt{\sum_{i=1}^{i=N}(x_i-\bar{x})^2} = \sqrt{(x_1-\bar{x})^2 + (x_2-\bar{x})^2 + \cdots + (x_N-\bar{x})^2}.$$

If we now regard this as an n-dimensional Euclidean hyperspace relationship, where each of the N points represents a "dimension" within a 1-dimensional distribution, we can see that $\sqrt{\Sigma(x-\bar{x})^2}$ is the total absolute distance of all the points $(x_1, x_2, x_3, ..., x_N)$ from the mean of all the points. We can further rearrange the expression for the standard deviation as

$$s = \sqrt{N} \times \left[\frac{\sqrt{\Sigma(x-\bar{x})^2}}{N} \right].$$

I've written it out like this to demonstrate that the term within the square brackets is the average absolute distance of all the points from the mean of all the points, which, when multiplied by $\sqrt{N}$, gives the standard deviation.

Thus, the term "deviation" in the definition of standard deviation is a measure of the absolute distance of all points from the mean of those points, irrespective of whether they are less than or greater than the mean.

2.4 Probability

Probability is all around us from conception to the grave. It is familiar to everyone and used in everyday life and language. And yet by its very nature it has a somewhat ephemeral ring to it, and its definition does not easily trip off the tongue. The Chambers Twentieth Century Dictionary (1976) doesn't help very much, where we find "quality of being probable" and "that which is probable", neither of which is very enlightening. But next there is "chance or likelihood of something happening", which looks more hopeful. When we turn to the definition of *chance* we have "that which falls out or happens fortuitously, or without assignable cause": "an unexpected event": "possibility of something happening": "probability". We seem to be going 'round in circles, so let's try *likelihood* where we find "similitude": "semblance": "resemblance": "probability"; we *are* going 'round in circles.

Perhaps we should start with intuitive logic and something we can really hang our hats on. Death is the only certain outcome of life. Indeed, there are those jaded individuals who regard life merely as a terminal disease, the ultimate outcome of which has a p-value of unity. On the other hand the fact that an individual exists as that particular entity must have a relatively low p-value, which probably (it's that word creeping in again) tends to zero but is clearly not zero because of that particular existence. I say a relatively low p-value because that particular ovum (one of about 400 released at random) had to have been fertilised by that particular sperm (one of legion which is also random) to produce that particular individual. A different ovum or a different sperm would have produced a different individual.

In the last example p was low, but some things have a probability of zero, as they are clearly impossible. I well remember the first time in my life I was perplexed. My doting Grandmother would read me the following nursery rhyme:

> Hey diddle diddle, the cat and the fiddle
> The cow jumped over the moon
> The little dog laughed to see such fun
> And the dish ran away with the spoon.

Apparently, I was about 2½ years old and wasn't too concerned about the cat or the fiddle or about the cow jumping over the moon. At that stage I wasn't too conversant with Newton's laws of motion, gravity, propulsion or the distance between the earth and the moon, though clearly here $p = 0$. I liked the idea of the little dog laughing as the lady next door had a dog that laughed, but I just could not cope with the dish running away with the spoon. Even at that early age I argued that this was a recognisable impossibility ($p = 0$) as neither the dish nor the spoon had legs. Thus, probability has a range from 0.0, absolute impossibility, to 1.0, absolute certainty.

Very few events occur with probabilities of 0.0 or 1.0; most fall somewhere in the middle, some closer to 0.0, others closer to 1.0. It will not have gone unnoticed that there are two fundamental types of humans, male and female. There are other subclassifications, but these are of no importance. Contrary to popular belief the two varieties are very similar. The most readily appreciated differences are anatomical but these are just quantitative, not qualitative. Embryologically they are identical, and there isn't anything I've got (I happen to be male) that the next female I bump into hasn't got and vice versa. It is also obvious that the two types are about equally prevalent. Each time a new human is conceived there is a 50% chance that it will be female. It can't be both (or not very often) so there is an equal chance that it will be male. Thus, the probability either of being female or being male is 0.5.

This also applies when we spin a coin – it will end up either heads or tails. This assumes, of course, that we have a regular coin with both a head and a tail and that it did not come to rest on its edge, which is not impossible but extremely unlikely. The chance of the latter occurrence can be calculated with a number of assumptions about the angular velocity and momentum of the spin, the thickness of the edge compared with the radius, the elasticity of the coin (remember it's going to bounce) and a number of other things. Just for fun I did this calculation for a U.K. 2p coin and got a probability of 10^{-14}. (Extraordinary what some people do for fun isn't it, but I've been in medicine for some time now and *nothing* surprises me!) This result is probably not in error by more than three orders of magnitude either way. However, if 10^{-14} is correct and if the coin were spun once every 4.75 seconds, then you could expect it to land and come to rest on its edge just 1000 times in the whole lifetime of the universe to date (15,000,000,000 years). This is what is meant by the chance of an occurrence being vanishingly small. Occasionally you just have to trust the mathematics without experimental verification, unless of course you have phenomenal stamina and even greater stupidity. We know that the result will be either a head or a tail, no other option is open. Therefore, the number of times tails is observed must be the number of times heads was not observed, which is given by the number of spins minus the number of heads. If a large number of spins is carried out then the frequencies of heads and tails will both tend to 0.5 as, for each spin, the probability of the outcome being heads is equal to the probability of the outcome being tails.

Similarly, if you roll a die there is a probability of ⅙ that any given number from 1 to 6 will lie face upwards when it comes to rest. You will immediately jump to the conclusion that the die had six faces; however, if it had 12 faces (I've never seen one like this but it could exit) there would be a probability of ¹⁄₁₂ of showing a given number.

In these various examples the probability of an event occurring can be defined by the system in question, because a priori we can define the probabilities from the intrinsic properties of the system. Probability in other types of systems cannot be so defined. An example is the number of deaths per passenger mile in air travel. This is "empirical" probability dictated by the experience of actual events as, before the aeroplane was invented, there were no casualties due to this agency. In such systems all we can do is count up the total number of people killed and divide this by the total number of passenger miles travelled. I do this quite frequently as I am flying-phobic, and I find it quite reassuring because in these terms air travel is even safer than walking down a busy street of an average small town like Cambridge. This, obviously, is quite illogical (but that's part of the definition of a phobia) because what I really want to know is whether I am going to be killed in the next aeroplane I step into. The overall probability cannot tell me this, as these are random events which apply to populations and cannot be applied to a specific individual who is part of those populations.

Probability, which is also defined by experience and experiment, occurs in all measurement systems. In the previous section I intimated that two independent measurements of the distance between two points marked on a sheet of paper using the same ruler may not be identical. This is due to an observer and measurement system error. In a number of trials we may not place the zero mark exactly adjacent to the first point each time. Likewise, we may not read off the scale mark adjacent to the second point correctly each time. Indeed, the calibration on the ruler may not allow us to do this with the precision we would wish. For instance, if the two points are just under 2 inches apart and the ruler is marked with gradations of ⅛ inch, we will be able to see that the point is not adjacent to either the 1⅞ or the 2 mark. We will be able to see that it may be nearer the 2 than the 1⅞ mark and we would have to estimate, by eye, how much nearer the 2 mark it is. However, we could do better than this by using a ruler which is marked out in $\frac{1}{16}$- or better still $\frac{1}{32}$-inch divisions. This, of course, is increasing the resolution of the system. Nevertheless, we still may not be able to read off an exact distance, particularly if the "true" distance between the points is something tricky like $1^{67}/_{71}$ inches, which it could well be. The problem here is that I've never met a ruler marked off in 71ths of an inch. All we could do in this particular instance is repeat the measurement a number of times and take the arithmetic mean. The under and over estimates are likely to cancel out and we will end up with a "best estimate" or at least a reasonable estimate of the true distance. Each time we make a measurement there will be a probability that that particular measurement will have some error associated with it.

Probability and the ways in which it arises was introduced here to remind you of some of the basic concepts and, needless to say, it will be cropping up again later.

2.5   Distributions

There are many types of distributions, but the three most commonly encountered in flow cytometry (as well as most other disciplines) are the Gaussian binomial and Poisson distributions. The distribution depicted in Figure 2.1 would be somewhat difficult to handle; it is not symmetrical and so it might be difficult to find a continuous mathematical function to describe it. However, help is at hand with a mathematical transform.

2.5.1   Mathematical transforms

Most biologists throw up their hands in horror at the mere mention of a mathematical transform, and before we embark upon this for the distribution in Figure 2.1 we will look at a very familiar, but not generally appreciated, example. The definition of pH is the log of the reciprocal of the hydrogen ion concentration, which may be rendered in mathematical notation as

$$pH = Log_{10}(1/[H^+]).$$

Everyone who is likely to read this book must have measured pH at some time or another and felt quite happy doing so. But how many people really feel comfortable thinking in terms of reciprocals, let alone taking logs of those reciprocals? Many biologists would feel highly uncomfortable with this concept if they thought about it, and I suspect that many would find some difficulty calculating the hydrogen ion concentration from the previous equation. Nevertheless, you feel comfortable with the number which comes out of the pH meter even though this is a mathematical transform and not a particularly easy one to comprehend. Why is this? I would venture to suggest that you feel comfortable with the number because you are familiar with the pH meter, you have some perception of what the number represents and you have forgotten that it is a mathematical transform. If we "unscramble" the above equation to put [H⁺] on the left-hand side we get

$$[H^+] = 10^{-pH} = 1/10^{pH}.$$

We know that pH 7 is "neutral" which, in [H⁺] terms, equates to 10^{-7}, the hydrogen ion concentration of water, and that pH 1 is "acid" which gives [H⁺] = $10^{-1} = 0.1$. Thus, as [H⁺] increases so pH decreases and as [H⁺] decreases so pH increases, which is the basis for the reciprocal relationship expressed in the definition of pH. Needless to say this is a convention, there is nothing absolute about it, and it could just as easily have been defined the other way around in terms of [OH⁻]. It is also interesting to note that, strictly speaking, there is no ZERO on the scale, as there is always a chance that there will be at least one OH⁻ at one end of the spectrum and at least one H⁺ at the other. Conceptually, the nearest thing to an origin or ZERO in this measurement system is "neutral", which just happens to have a value of 7.

I hope I've convinced you there is nothing to be feared in making mathematical transforms. If you can cope with pH then you can cope with the other transforms in this book, and we can now proceed with a transform of the data shown in Figure 2.1. This is very much simpler than the pH example. All I have done is to take a sheet of semi-log paper, turn it on its side so that the X-axis (abscissa) has the log scale along it, and plot the frequencies on the linear Y-axis. This is shown in Figure 2.3, which is a log transform of the data, and we can now see that the distribution is almost perfectly symmetrical. However, when you do this you must remember to take the logarithm of the scale reading,

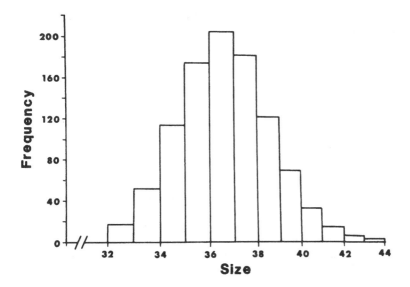

Figure 2.3. Log transform of the data in Figure 2.1 which makes the distribution symmetrical.

as this converts into linear units. This is necessary because the standard deviation is a linear variable. You don't even need semi-log paper for this exercise. All you have to do is take the log of the X-axis scale reading of Figure 2.1 and do the plots on regular linear graph paper.

2.5.2   Gaussian

The Gaussian or so-called normal distribution is anything but normal. It was originally formulated by de Moivre (1756), an English mathematician in spite of his French-sounding name. Gauss (1809) and Laplace (1812) also derived the distribution, which was found to describe inter-observer deviations of stellar positions in astronomical records. This is directly analogous with our problem of measuring the distances between points using a ruler, and it was from these types of findings that it obtained its third name, the error function. The formula for the bell-shaped normal distribution curve is

$$y = \frac{1}{\sigma\sqrt{2\pi}} e^{-(x-\bar{x})^2/2\sigma^2}.$$

This looks a bit horrific, and again we must consider the meaning of the various components in the equation. We've met $(x-\bar{x})^2$ for the fourth time now. We've also met with σ and σ^2 which leaves only the constants 2, e and π. e is the exponential number which, for what it's worth, has a value of about 2.71828, but where does π, which just happens to have a value of 3.14159 to five decimal places, come from? If we plot out values of y for given values of x and two different standard deviations, we obtain the curves shown in Figure 2.4. The y-values are zero at both $x = -\infty$ and $+\infty$. When $x = \bar{x}$ the exponential component is unity as $(x-\bar{x}) = 0$ and $e^0 = 1$, hence $y = 1/\sigma\sqrt{2\pi}$. If we now find the total areas under these curves (by integration) they come to unity, but only

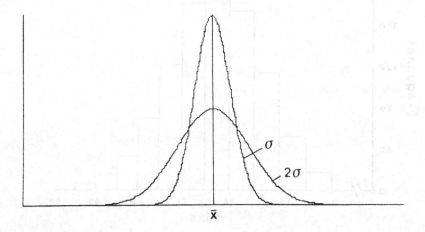

Figure 2.4. Two normal curves with different standard deviations σ, where heights and spreads are dependent on σ.

because of the constant $1/\sigma\sqrt{2\pi}$. If this component had been omitted then the areas under the curves would be $\sigma\sqrt{2\pi}$, so the constant $1/\sigma\sqrt{2\pi}$ is included in the equation to normalise the areas under the curves to unity.

There is one further symbol we haven't discussed yet, the minus sign $(-)$ in front of the exponent. When x is less than $\bar{x}$ the expression $(x - \bar{x})$ is negative; thus the whole exponent is interpreted as positive and the curve rises. When x is greater than $\bar{x}$ the expression $(x - \bar{x})$ is positive, so the exponent is interpreted as negative and the curve falls. Thus, the function must be symmetrical about the mean $\bar{x}$. You will also appreciate that if the expression $(x - \bar{x})$ had been written as $(\bar{x} - x)$ then there would have been no need for the minus sign in front of the exponent. Just as light in ray diagrams is always depicted as being propagated from left to right, so in mathematics we "start" with smaller numbers being depicted on the left with the larger ones towards the right. Again, this is merely convention, which in this instance dictates that a minus sign must be placed in front of the exponent.

Another fundamental concept now comes to mind. This concerns the negative set of numbers. Just for the hell of it, put two pens on this page. Now, take one away and you won't be surprised to find that only one remains. This really isn't as daft as you might be thinking. Now, put the subtracted pen back to restore the page's original complement of two. The next bit is a little tricky. One at a time take three pens away from the page. You are OK for the first two, but as soon as the page has zero pens you are in trouble – unless, of course, you imagine there is another pen on the page so that you *can* subtract it. Having done so, you could then express the operation by saying that the page possesses one "negative" pen, which, I'm sure you will appreciate, is very different from saying it doesn't possess a "positive" pen. Everyone has seen a "positive" pen, but has anyone ever seen a "negative" pen?

I've included this little anecdote to demonstrate that the negative range of numbers is totally abstract. They do not exist as "real" entities; nevertheless, they can be used. Furthermore, they are used so often that we have forgotten that they are purely abstract and in consequence they have taken on a mantle of reality. This idea has been with us almost since man emerged as a distinct species and it was one of our first journeys into the abstract. It embraces the concept of having something today, owing a debt and paying for it tomorrow. In the early days it was food now, in the West it's TV sets, washing machines, compact disk players, bigger cars and various other trivia. (It is also the unfortunate reality of government budget-deficit financing in the majority of the Western world – a process which, if unabated, can only end in disaster.)

The relationship just given for the normal curve is in a slightly inconvenient form, as the maximum height of the curve, $1/\sigma\sqrt{2\pi}$, will vary from curve to curve depending on standard deviations σ. This is circumvented by another transform where $(x - \bar{x})/\sigma$ is replaced by the symbol z, which reduces the normal curve to its more usual form:

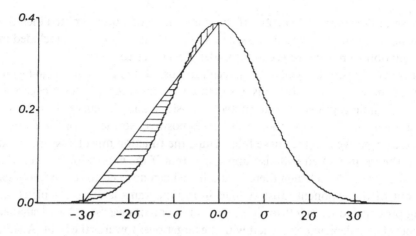

Figure 2.5. Standardized normal curve, where the standard deviation is unity.

$$y = \frac{1}{\sqrt{2\pi}} e^{-z^2/2}.$$

Now, what exactly has happened here? First, the equation is very much simpler, and simplification is always welcome, but σ seems to have disappeared and isn't this a bit of a "fiddle"? Well yes, it is a bit of a fiddle if you wish to regard it as such, but it is a perfectly legitimate fiddle in order to express the curve in a standard form that is applicable to all normal distributions. However, σ has not, in fact, disappeared; it now has a value of unity. The transform $z = (x - \bar{x})/\sigma$, hence $z^2 = (x - \bar{x})^2/\sigma^2$ so that the exponential component becomes $z^2/2$, now expresses the variable $(x - \bar{x})$ – in other words, the deviation from the mean – in terms of numbers of *unit* standard deviations from the mean, where $\sigma = 1.0$. The X-axis scale now reads from $-\infty$ to $+\infty$ with the mean always located in the middle at zero (i.e. 0.0). This enables the relationship to be used as a convenient probability function, shown in Figure 2.5 and discussed further in Section 3.1. Note that the maximum height of the curve is very nearly 0.4 which is $1/\sqrt{2\pi}$, $\bar{x}$ is set at 0.0 and the total area under the curve is unity. A very rough check on the latter can be carried out by drawing a line from -3σ to the peak height at $x = 0.0$ as shown. The area of this triangle is $0.4 \times \frac{1}{2} = 0.6$. Obviously, this is an overestimate of the area under the curve to the left of the origin, which should be 0.5. But the horizontal hatched area is included in the area of the triangle and this is not part of the area under the curve; this is larger than the vertical hatched area that is part of the area under the curve but is excluded from the area of the triangle.

2.5.3 Binomial

This distribution is concerned with the occurrences of mutually exclusive events and is described by the relationship

$$(p+q)^n = 1,$$

where p is the chance of something happening and therefore q must be the chance of that something not happening. The coefficient n is the number of times the event is observed or the number of trials carried out. It is best illustrated by pursuing the example of rolling dice. There is a probability of $p = \frac{1}{6}$ of obtaining a 5 when a single die is rolled; here, n is 1. We can now see that $q = \frac{5}{6}$, which is the probability of obtaining a number other than 5. The act of rolling a die for a first time and getting a 5 cannot possibly influence what will happen if you then roll it a second time, so the probability of getting a 5 on the next roll is also $\frac{1}{6}$. Thus, $n = 2$ in the preceding equation and on expansion we get

$$p^2 + 2pq + q^2 = 1,$$

where the probability of getting 5 twice in succession, which is the same as rolling two dice simultaneously, is $p^2 = (\frac{1}{6}) \times (\frac{1}{6})$ or $\frac{1}{36}$. The probability that *one,* but not the other, die will show a 5 is given by $2pq = 2 \times (\frac{1}{6}) \times (\frac{5}{6}) = \frac{10}{36}$. This leaves the probability q^2, which is $(\frac{5}{6}) \times (\frac{5}{6}) = \frac{25}{36}$, that neither die will show a 5. The divisor (the number underneath in the fraction) is the same, 36, for each of these probabilities and the numerators (the numbers on the top) are 1, 10 and 25 which, nice and neatly, add up to 36; thus, the total probability of getting some combination of numbers is unity.

If we now extend this to a throw of three dice simultaneously, the binomial expansion becomes

$$p^3 + 3p^2q + 3pq^2 + q^3 = 1,$$

and again $p = \frac{1}{6}$ for a 5 and $q = \frac{5}{6}$ for any number other than 5. The calculation of the probabilities for all possible outcomes is shown in Table 1.

In these examples the probability of getting an given number remained constant from trial to trial. We should now consider what happens if the probability of an event occurring is modulated by the event. Perhaps the best example is drawing lots. If we start off with a black bag (so that you cannot see into it) containing 50 white chips and 50 blue chips, there is a $\frac{50}{100}$ (i.e. 0.5) probability

Table 1. *Binomial probabilities for $n = 3$*

Binomial term	Outcome	Probability
p^3	3 fives	$\frac{1}{6} \times \frac{1}{6} \times \frac{1}{6} = \frac{1}{216}$
$3p^2q$	2 fives, 1 other	$3 \times (\frac{1}{6} \times \frac{1}{6}) \times \frac{5}{6} = \frac{15}{216}$
$3pq^2$	1 five, 2 others	$3 \times \frac{1}{6} \times (\frac{5}{6} \times \frac{5}{6}) = \frac{75}{216}$
q^3	3 others	$\frac{5}{6} \times \frac{5}{6} \times \frac{5}{6} = \frac{125}{216}$

that the first chip you pull out of the bag will be blue. What, now, is the probability of getting another blue chip on the second draw? There remain 49 blue chips in the bag out of a total of 99. Thus, the new probability of drawing a second blue chip is $^{49}/_{99}$, or 0.494949, which is slightly less than on the first draw. Thus, the probability that the first two chips drawn from the bag are blue can be calculated as $(^{50}/_{100}) \times (^{49}/_{99})$ or 0.5×0.494949. We now have 48 blue and 50 white chips remaining, hence the probability that the next chip drawn will be blue is $^{48}/_{98} = 0.489796$. Therefore, the probability that the first three chips drawn will all be blue is given by $(^{50}/_{100}) \times (^{49}/_{99}) \times (^{48}/_{98})$.

You will recognise that these types of concepts are encountered in cell sorting purity. A cell of interest either will or will not be in a sorted droplet, it can't be half in, which also implies that it can't be half out. Moreover, if coincidence is occurring then the probabilities of getting 2 wanted cells or 1 wanted and 1 unwanted are governed by the binomial distribution.

2.5.4   Poisson

A different type of analysis problem arises if we are standing by a road observing cars on the near side passing from right to left (we are in England). We can count how many cars pass in 10 minutes, or in any other interval of time, hence we can calculate a flow rate. It is self-evident, however, that it would be absolutely ridiculous to try to count how many cars *did not* pass the same point in the same time interval, and events involving these types of observations cannot be handled by the binomial distribution. Cars passing a point along a road are isolated events occurring in a continuum of time, as are cells passing the analysis point in a flow cytometer, and events of this nature are described by the Poisson distribution (Poisson 1837). I will not go into the mathematical derivation of this distribution, although it is very straightforward. If you want full details read Chapter 8, "Goals, floods and horse kicks – the Poisson distribution", in *Facts from Figures* by Moroney (1990, first published in 1951 and still in print).

In order to use the Poisson distribution all we need is z, defined as the average number of times an event occurs within a continuum. The latter could be length, time, volume, area or anything, and the probability of observing the event zero times or once, twice, thrice, etc. within a defined (and constant) increment of that continuum is given by the successive expansion terms of the expression $e^z \times e^{-z}$. The mathematical notation of each element $p(n)$ of the Poisson distribution is $p(n) = z^n e^{-z}/n!$, where $p(n)$ is the probability of n events being observed. The value of n can be anything from 0 to ∞, and $n!$ is factorial n. The notation for the summation for the whole distribution is

$$P = \sum_{n=0}^{n=\infty} \frac{z^n e^{-z}}{n!},$$

which has unit value and is a pretty fancy way of writing 1. In most cases that are likely to be encountered in practice $p(n)$ becomes vanishingly small for

Table 2. *Poisson distribution*
terms up to and including $n = 4$

Frequency that the number of events, n, are observed	Probability of observing the n events
0	$z^0 e^{-z}/0!$
1	$z^1 e^{-z}/1!$
2	$z^2 e^{-z}/2!$
3	$z^3 e^{-z}/3!$
4	$z^4 e^{-z}/4!$

values of n greater than 30 or so, as factorial 30, the divisor (that's the bit underneath in the fraction), is a very large number. Values for n greater than 4 should never occur in flow cytometry; the terms of the Poisson distribution up to and including 4 are given in Table 2. The first term, $z^0 e^{-z}/0!$, reduces to e^{-z} as z^0 and 0! are both unity, and the second term reduces to ze^{-z} as z^1 is z and 1! is unity.

3

Probability functions

Descriptions of the three distributions to be considered in this chapter were given in Chapter 2, where in each case the total "area" of each function was unity. In order to use any formula as a probability density function it is necessary that the total area of the function be unity, since probability has a range from 0.0 to 1.0 and it is the area under these various functions that is proportional to probability.

3.1   Gaussian distribution

You will recall from Section 2.5.2 that the normal curve had to be mathematically "manipulated" by the factor $1/\sigma\sqrt{2\pi}$ to ensure that the total area under the curve be unity. Moreover, it was further manipulated with the z-transformation, where $z = (x - \bar{x})/\sigma$, so that the mean $\bar{x}$ always has a value of zero and the standard deviation σ is unity, producing the standardised function. The formula is reproduced here so you don't have to thumb back to Section 2.5.2:

$$y = \frac{1}{\sqrt{2\pi}} e^{-z^2/2}.$$

3.1.1   Cumulative frequency

We can now compute from this equation values of y for given values of z, which has units of numbers of standard deviations, together with the cumulative frequencies. The latter represent the area under the normal curve up to and including each particular value of z. This is carried out by a numerical integration routine (erf(z); see Appendix 1) as our equation is not integrable by regular analytical methods. These data are shown in Table 3 for values of z ranging from -4.0 to $+4.0$ in steps of 0.5, where it will be noted that the maximum frequency, which occurs at $z = 0.0$ (i.e. $\bar{x}$), is 0.3989 ($1/\sqrt{2\pi}$). Also, it will be noted that the cumulative frequency is 0.0 at $z = -4.0$, 0.5 at $z = 0.0$ and 1.0 at $z = +4.0$. Thus, half the area of the curve lies to either side of the mean.

Table 3. *Frequencies and cumulative frequencies for the normal distribution from* $\sigma = -4$ *to* $\sigma = +4$ *at intervals of* 0.5σ

Standard deviation	Frequency	Cumulative frequency
−4.00	0.0001	0.0000
−3.50	0.0009	0.0002
−3.00	0.0044	0.0013
−2.50	0.0175	0.0062
−2.00	0.0540	0.0228
−1.50	0.1295	0.0668
−1.00	0.2420	0.1587
−0.50	0.3521	0.3085
0.00	0.3989	0.5000
0.50	0.3521	0.6915
1.00	0.2420	0.8413
1.50	0.1295	0.9332
2.00	0.0540	0.9772
2.50	0.0175	0.9938
3.00	0.0044	0.9987
3.50	0.0009	0.9998
4.00	0.0001	1.0000

However, we should note that the values of 0.0 at $z = -4.0$ and 1.0 at $z = +4.0$ are approximations since the cumulative frequency strictly only reaches 0.0 and 1.0 at $-\infty$ and $+\infty$ respectively, but in reality the vast majority of the curve is contained within these limits.

The cumulative frequency distribution represents a convenient method of calculating the area that lies between two limits. If we take the interval −0.5 to 0.5, we find cumulative frequencies of 0.3085 and 0.6915 (respectively) and the difference is 0.383. If we now translate this into population terms, where the latter is normally distributed, we have 38.3% of the total population lying within the interval of ±0.5 standard deviations about the mean. Similarly, at the ±1 standard deviation limit about the mean we have 0.8413 − 0.1587 which translates to 68.26%, and at ±1.5 standard deviations we have 0.9332 − 0.0668 which is 86.7%. This is represented graphically in Figure 3.1. A table of normal distribution probabilities for two-decimal-place standard deviation intervals is given in Appendix 2.

Let us now suppose we have two normal distributions with identical standard deviations whose means are separated by 3 standard deviations, which gives rise to some overlap. This is shown on Figure 3.2, which might represent

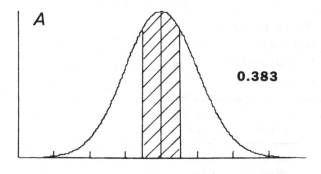

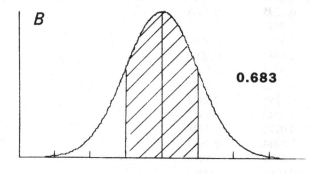

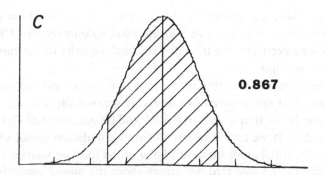

Figure 3.1. Graphical illustration of integrating between limits on the normal curve.

the distributions of two sets of recorded observations of the positions of two stars in very close apposition. If another measurement is now made and it falls in the position marked by the arrow, we must decide to which star this new measurement belongs. This new measurement lies within the interval of +0.5 to +1.0 standard deviations from the mean position of the first star and −2.5

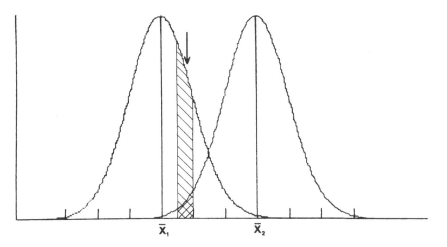

Figure 3.2. Two overlapping normal distributions showing the mean and standard deviation of each, with the arrow marking the position of a "new" data point.

to -2.0 standard deviations from the mean position of the second star. We can see from the cumulative frequency distribution that $+0.5$ and $+1.0$ are associated with values of 0.6915 and 0.8413 respectively, a difference of 0.1498, and that -2.5 and -2.0 are associated with values of 0.0062 and 0.0228, a difference of 0.0166. These two numbers, 0.1498 and 0.0166, represent the probabilities that the new observation belongs to star 1 and star 2, respectively. We would then conclude that it was considerably more likely, almost tenfold as $(0.1498/0.0166) = 9.02$, that the new observation belongs to star 1. However, it is not impossible, just tenfold less likely, that it could belong to star 2.

3.1.2   Population sampling

The type of specific question asked in the previous section is not one that will arise in flow cytometry data analysis, where we will wish to know if two distributions are different. However, the principles in the latter are essentially identical to those in the former, where the probability of difference will depend on separation of means and dispersion about the means.

However, before we embark upon this in the next chapter, it is necessary to appreciate some subtle distinctions between analysing the differences between two *whole* populations and differences between two samples drawn one from each of those two populations. This is best illustrated with an example of samples drawn from a single population. I asked a random-number generator to pick five numbers from a normal distribution containing 10,000 values (cells), with $\bar{x} = 0.0$ and a standard deviation of 200.0. The numbers, represented by A, B, C, D and E, had values of -343, -137, -61, 176 and 443, respectively. If we now divide these by 200 to convert to standard deviation units we get

Table 4. *All possible combinations of five variables*

	Sample size				
	2	3	4	5	x
A	+ + + +	+ + + + + +	+ + +	+	−1.7
B	+ + + +	+ + + + + +	+ + + +	+	−0.7
C	+ + + +	+ + + + + +	+ + + +	+	−0.3
D	+ + + +	+ + + + + +	+ + + +	+	0.9
E	+ + + +	+ + + + + +	+ + + +	+	2.2

−1.7, −0.7, −0.3, 0.9 and 2.2, respectively (rounding to one decimal place). These were used in various combinations to represent samples drawn from the parent population. We can have ten combinations of two at a time, such as $A+B$, $A+C$, $A+D$ etc.; there are also ten combinations of three at a time, such as $A+B+C$, $A+B+D$, $A+B+E$ etc. Five combinations of four are possible (where one of the five numbers is omitted from each combination), and finally there is one combination of five. These are summarised in Table 4 together with the values of each number (x column).

The means and standard deviations of the combinations appearing in Table 4 were calculated, the latter as $s = \sqrt{\Sigma(x-\bar{x})^2/N}$, where s represents the sample standard deviation (to distinguish it from the standard deviation σ of the whole population from which it was derived). These standard deviations for each sample combination are plotted on the ordinate of Figure 3.3 for the number of values in the sample on the abscissa. Two observations are important regarding this figure. First, the ranges of the sample standard deviations decrease as the number in the sample increases, and second, $\bar{x}$ and s for the sample of five were 0.08 and 1.35, respectively. This value for $\bar{x}$ is very close to that of the parent population, 0.0, but the standard deviation s seems to be considerably greater than unity $= \sigma$ of the parent population, and we will see in the next chapter how we can assess this discrepancy.

3.1.3 Standard error of the mean

The data in Figure 3.3 also show that we need not only the sample means and standard deviations to assess differences between sample distributions, but also the number of "events" (cells), as the sample standard deviation is dependent on the number in the sample. The sample standard deviation is inversely proportional to $\sqrt{N}$, where N is the number in the sample, and this gives the relationship $s/\sqrt{N}$ which is the standard error of the mean, $SE_{\bar{x}}$. The relationship between the standard deviation and standard error of the mean can readily be appreciated from the third expression given for the standard deviation in Section 2.3.3:

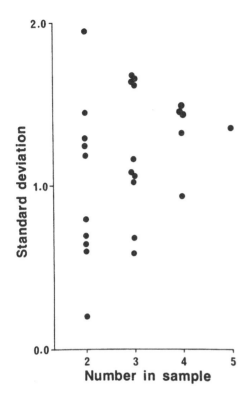

Figure 3.3. Standard deviation of a sample drawn from a population plotted against the sample size.

$$s = \sqrt{N} \times [\sqrt{\Sigma(x-\bar{x})^2}/N].$$

If we now divide this expression by $\sqrt{N}$ we get the standard error of the mean, where

$$SE_{\bar{x}} = \sqrt{\Sigma(x-\bar{x})^2}/N.$$

Thus, the standard error of the mean is the average of the sum of the absolute distances of all the points from the mean of all the points, and it is this parameter that is used in assessing differences between *samples* derived from distributions.

3.1.4 Standard error of the standard deviation

A similar parameter can be defined for the standard deviation: the standard error of the standard deviation, SE_s, where $SE_s = s/\sqrt{(2 \times N)}$. Expressing this in the same notation as above, we obtain

$$SE_s = \sqrt{\Sigma(x-\bar{x})^2}/(N \times \sqrt{2}),$$

which is the average of the sum of the absolute distances of all the points from the mean of all the points, divided by $\sqrt{2}$.

3.1.5   Skew and kurtosis

These two parameters can sometimes be useful for deciding how well a given distribution approximates to the Gaussian. If a distribution is perfectly normal, its skew and kurtosis will have values of 0.0 and 3.0, respectively. These parameters are calculated from the so-called second, third and fourth moments about the mean, where

$$\text{skew} = \frac{m_3}{m_2 \sqrt{m_2}},$$

$$\text{kurtosis} = \frac{m_4}{(m_2)^2},$$

and the formulae for these moments are as follows:

$$m_2 = \left(\frac{1}{n} \sum_{i=1}^{i=n} x_i^2\right) - (\bar{x})^2,$$

$$m_3 = \left(\frac{1}{n} \sum_{i=1}^{i=n} x_i^3\right) - \left(\frac{3\bar{x}}{n} \sum_{i=1}^{i=n} x_i^2\right) + 2(\bar{x})^3,$$

$$m_4 = \left(\frac{1}{n} \sum_{i=1}^{i=n} x_i^4\right) - \left(\frac{4\bar{x}}{n} \sum_{i=1}^{i=n} x_i^3\right) + \left(\frac{6(\bar{x})^2}{n} \sum_{i=1}^{i=n} x_i^2\right) - 3(\bar{x})^4.$$

The second moment, m_2, is the sample variance, which is also written in its more compact form as

$$m_2 = s^2 = \frac{\sum x^2}{n} - \bar{x}^2.$$

3.2   Poisson distribution

Let us suppose that we wish to calculate the probability of coincidence in cell sorting, and that the object we wish to collect is the y-chromosome. This is present at a frequency of $1:46$ and we want a total of 10^7 in order to construct a genomic library. Let us also suppose that we apply the argument that, as this is the rarest of the chromosomes, we will employ a three-droplet sort to minimize the chance of missing the object of interest. We are using a regular cell sorter with a through-put rate of 5000 chromosomes per second and a droplet formation frequency of 45,000 per second. This is the total quantity of information required to calculate the purity of the final product and the time it will take to collect the requested number of chromosomes.

Statistics governing objects flowing through a cytometer is a Poisson problem, just as cars passing a fixed point on a road is a Poisson problem. If 5000 objects are flowing through the instrument per second and the droplet formation frequency is 45,000 times per second, then "on average" one droplet in nine will contain an object. However, as we are sorting three droplets at a time,

Table 5. *Poisson probabilities for zero to four chromosomes being in a given set of three sorted droplets*

Objects per sorted cohort of three droplets	Poisson probability
0	0.71892
1	0.23725
2	0.03915
3	0.00430
4	0.00036

on average one in every three sorted cohorts of three droplets will contain an object. Thus, the expectancy z to be plugged into the terms of the Poisson distribution expansion is $\frac{1}{3} = 0.333$. The Poisson probabilities of obtaining 0, 1, 2, 3 and 4 chromosomes per set of three droplets are shown in Table 5.

We can see from Table 5 that about 72% of any given set of three droplets will contain no chromosomes, about 23% will contain one and almost 4% will contain two. However, we will not be triggering the electronics (hopefully) if there are no events at all in a given set of three, and consequently we must ignore the first term in Table 5. Thus, the probability of having at least one object in a given set of three droplets is $1.0 - 0.71892$, which is 0.28108. We can now correct the numbers in Table 5 so that various probabilities become relative to 0.28108; this is shown in Table 6, where the probabilities have been converted to percentages of 0.28108. Thus, 84.4% of the cohorts of three droplets will contain one chromosome, 13.9% will contain two, 1.6% will contain three and 0.1% will contain four. Two further factors must be appreciated. First, we are

Table 6. *Probabilities in Table 5 related to the single-event probability of 0.28108*

Objects per sorted cohort of three droplets	Poisson probability (%)
1	84.4
2	13.9
3	1.6
4	0.1

referring here to all chromosomes, not just to the y that we require; second, if there is coincidence of two or more chromosomes then at least one of these may or may not be the object of interest. This is a binomial problem super-imposed on the Poisson probability, which is developed further in the next section.

3.3   Binomial distribution

We have seen in Section 2.5.3 how the binomial distribution was used to calculate the probabilities for occurrences of given numbers and combinations thereof when throwing dice. We are now confronted with exactly the same problem in our chromosome sorting. Any object can either be wanted or un-wanted in a droplet that is being sorted. The probability of a given object being with a given droplet is governed by Poisson statistics, as shown in the last section. If two chromosomes are in a sorted set of three droplets then they could be two wanted (i.e. y-chromosomes in this particular example) or one wanted plus one unwanted. However, we cannot have two unwanted, as the instrument was not programmed to sort this combination; there must be at least one wanted chromosome to trigger the sorting logic. If three chromosomes are in a sorted cohort of three droplets then we could have three wanted, two wanted plus one unwanted or one wanted plus two unwanted. Again, we cannot have three unwanted.

The binomial expansion for three simultaneous events has already been considered (see Section 2.5.3 and Table 1). If we now take $n = 4$, the expansion becomes

$$p^4 + 4p^3q + 6p^2q^2 + 4pq^3 + q^4 = 1,$$

which, in the extreme example, would represent four chromosomes being sorted simultaneously. We could then have the following combinations: four wanted; three wanted plus one unwanted; two wanted plus two unwanted; one wanted plus three unwanted; but we could not have four unwanted. The probabilities of these occurrences for $p = 0.021739$, representing the fraction $1/46$ of wanted chromosomes in the population, are given in Table 7.

The total probability in Table 7 sums to 8.41×10^{-2}; the q^4-term, which is $(45/46)^4 = 9.16 \times 10^{-1}$, is not applicable because it represents only unwanted chromosomes which are not sorted. The probabilities for the occurrence of each combination can now be calculated relative to the total, 8.41×10^{-2}. These relative probability values are then multiplied by the number of expected wanted chromosomes to give their frequencies. These data are shown in Table 8.

The total frequency in Table 8, 1.035356, is the number of wanted events per coincident four sorted, which yields a value of 0.259. Similar calculations can be carried out for coincidence of both three and two events, yielding values of 0.3406 and 0.5055, respectively. Table 9 shows the Poisson probability for N, the number of coincident events sorted simultaneously, the total number of

Table 7. *Binomial probabilities of getting the various combinations of four chromosomes in a given set of three droplets*

Term	Combination	Probability
p^4	$4p$	$(\frac{1}{46})^4 = 2.23 \times 10^{-7}$
$4p^3q$	$3p+1q$	$4 \times (\frac{1}{46})^3 \times \frac{45}{46} = 4.02 \times 10^{-5}$
$6p^2q^2$	$2p+2q$	$6 \times (\frac{1}{46})^2 \times (\frac{45}{46})^2 = 2.71 \times 10^{-3}$
$4pq^3$	$1p+3q$	$4 \times (\frac{1}{46}) \times (\frac{45}{46})^3 = 8.14 \times 10^{-2}$
q^4	$[4q$	$(\frac{45}{46})^4 = 9.16 \times 10^{-1}]^a$

[a] Not applicable.

Table 8. *Absolute and relative binomial probabilities*

Term	Absolute probability	Relative probability	Frequency
p^4	2.23×10^{-7}	0.00000274	0.000011
$4p^3q$	4.02×10^{-5}	0.00049355	0.001481
$6p^2q^2$	2.71×10^{-3}	0.03327194	0.066544
$4pq^3$	8.14×10^{-2}	0.96732026	0.967320
Total	8.41×10^{-2}	1.00000000	1.035356

Note: The total of the relative binomial probabilities has been normalised to unity with frequency of wanted events.

Table 9. *Poisson probabilities, total numbers sorted, fractions of wanted events and wanted events sorted*

N	Poisson probability	Total number sorted	Fraction of wanted events	Wanted events sorted
1	0.844	84.4	1.000	84.4
2	0.139	27.8	0.506	14.067
3	0.016	4.8	0.341	1.637
4	0.001	0.4	0.259	0.104
Total	1.000	117.4		100.208

events sorted relative to a nominal 100, the fraction of events within each sorted group that are wanted and the number of events sorted that are wanted.

We can now see that the proportion of y-chromosomes in relation to the total number sorted is $100.208/117.4 = 85.26\%$, which is marginally higher than the percentage of single events sorted, 84.4%. This small difference of 1% is due to the relative proportions of binomially distributed wanted and unwanted chromosomes during coincidental sorting. In this particular application we would obtain about 110 y-chromosomes per second with 85% purity, and this would require about 25 hours of continuous sorting to obtain the target total of 10^7.

The take-home messages of the last two sections are that standard cell sorters are very efficient for sorting small numbers of cells with high precision and purity for identification purposes. However, they are not very efficient for sorting large numbers of cells for preparative work. Furthermore, purity must often be sacrificed to obtain a good yield (large numbers), but this problem can sometimes be overcome by carrying out a bulk pre-sorting enrichment procedure.

4

Significance testing and fit criteria

I've managed to get a significant proportion of this book written without embarking on statistics. This was intentional, as I appreciate that biologists generally "switch off" at its mention. However, we now have to bite the bullet. Statistics languish under many guises, and many people are instinctively sceptical particularly of the type generated by the advertising trade, which may boast that 8 out of 10 highly respected and influential individuals use ANED. The inference is that you too, who are probably not as highly respected or influential as the worthy characters in the advert, should use ANED so that you may become more highly respected and influential. We should ask just two questions. First, why do the manufacturers need to waste their money on advertising if 8 out of 10 of all people use the product, which begs the second question: 8 out of 10 of *which* people? The answer, of course, is that very few of the total number of people use ANED, so the manufacturers need to boost sales to satisfy the share-holders by increasing profits. The advertisers are very careful these days not to be seen perpetrating a blatant untruth, and so carefully target their research to the most likely users of ANED before they do the survey. It is true that 8 out of the 10 people *they interviewed* use ANED, but these people were not selected at random. (Incidentally, ANED stands for Automatic Nose Evacuation Device; the problem of ANED advertisers is that most people don't go around picking their noses, and those that do use a finger.)

Now, to get back to more sensible things. The two questions most commonly asked in flow cytometric data analysis are:

(1) Is there more than one population?
(2) If there is more than one, how many cells are in each?

In the present day and age, with the degree of computer sophistication we have available, this is far too naive an approach. We should restructure our thinking to include the following questions:

(1) Is there more than one population?
(2) If there is more than one, how far "separated" are they?
(3) What is the significance of that separation?

(4) If the populations are significantly separated then what are the estimates of the relative proportions of cells in each?

(5) What significance can we attach to the estimated proportions?

If we take immunofluorescence data as an example, we find results presented as "this sample contained 15.3% positive cells". Usually, the data are analysed by placing a vertical delimiter 2.5 standard deviations to the right of the mean of the control and scoring all cells to the right of this in subsequent test samples as positive. We will see later that this can produce so-called answers that may be in error by an order of magnitude. Moreover, when results are presented in this manner we do not know either how far apart the "negative" and "positive" might be or the dispersion in either population. Hence, the informed reader has absolutely no information which would enable an objective assessment to be made regarding the validity of the assertion that 15.3% of the population is positive. This is as bad, if not worse, than the ANED advertisers because therapeutic decisions may be based on the results. I submit that we should present results in the following type of format: "The analysis was compatible with two populations whose means were separated by a factor of 1.75, where the predicted positive fraction constituted 15.3% of the total at the 95% confidence interval". We can now get a feel for how different the populations actually were, and obviously we would be even more confident about interpreting the result of 15.3% positive if this had been obtained at the 99% confidence interval.

In order to present results in this manner we need to know the mean and variance of each population and then perform significance testing. Methods of analysis to obtain the mean and variance of the various populations in such data sets will be described in Chapter 7, but first we consider in this chapter the various significance tests that constitute the tools required to effect this approach.

4.1   Parametric

Parametric statistical tests are those based on some form of mathematical probability function. The most commonly used function is the normal distribution defined by the two parameters, the mean and standard deviation (which we have met previously). It goes without saying – though it is not often explicitly stated when they are used – that these types of tests apply only if the data have been shown to conform to the distribution being used to generate the statistic.

4.1.1   Standard error of difference

A central axiom in statistical theory is that the variance of the sum or difference of two independent and non-correlated random variables is equal to the sum of their variances. The standard deviation is *not* additive. This is a consequence of the definition of variance which, just to remind you, is given by

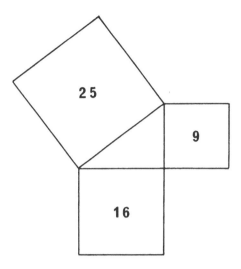

Figure 4.1. Classical 3-4-5 Pythagorean right-angle triangle with squares adjacent to the respective sides.

$$\sigma^2 = \frac{\Sigma(x - \bar{x})^2}{N}.$$

The standard deviation is the square root of the variance, and the relationship between these two parameters is most easily appreciated by considering the classical Pythagorean 3-4-5 right-angle triangle shown in Figure 4.1. The squares on the sides of the triangle are drawn adjacent to each side and the areas of these are shown, that is 9, 16 and 25 respectively. The length of the hypotenuse is obtained by summing 9 and 16 and taking the square root, which gives 5. The variance, a squared commodity, is analogous to the squares on the sides of the triangle, and the standard deviation is analogous to the lengths of each side. Hence, just as we sum the squares and not the lengths of the two shorter sides of the right-angle triangle, so we sum variances. It is the variance which is additive, not the standard deviation.

Whenever we carry out a sampling procedure we can never include the whole population, and we saw in Section 3.1.2 that the standard deviation s of a sample drawn from a population decreases with sample size. Moreover, the decrease in the standard deviation is proportional to $\sqrt{N}$, where N is the number of items in the sample, which gives rise to the standard error of the mean, $SE_{\bar{x}}$, where $SE_{\bar{x}} = s/\sqrt{N}$. If we now square this relationship we obtain the variance Var for a sample containing N items, where

$$\text{Var} = s^2/N.$$

Extending this to two distributions – with $\bar{x}_1, s_1, n_1$ and $\bar{x}_2, s_2, n_2$ representing (respectively) the mean, standard deviation, and number of items in each – we obtain the sum of the variance, $\mathrm{Var}(\bar{x}_1 + \bar{x}_2)$, where

$$\mathrm{Var}(\bar{x}_1 + \bar{x}_2) = \frac{s_1^2}{n_1} + \frac{s_2^2}{n_2}.$$

By taking the square root of this expression we obtain the standard error of difference between the means of the two samples. We know the difference between means, $|\bar{x}_1 - \bar{x}_2|$ or $\sqrt{(\bar{x}_1 - \bar{x}_2)^2}$, hence we can calculate the number of "standardized" standard deviation *units* that the difference represents by dividing the difference by the standard error of difference. This yields

$$\frac{\sqrt{(\bar{x}_1 - \bar{x}_2)^2}}{\sqrt{s_1^2/n_1 + s_2^2/n_2}},$$

where we have met the expression $\sqrt{(\bar{x}_1 - \bar{x}_2)^2}$ before as the absolute distance between points, in this case the distance between means. As an example, we may find that the above relationship gives a value of 1.5. Thus, the difference between the two means is only 1.5 standard errors of difference. Looking to the cumulative frequency distribution of the normal curve shown in Table 3, we find that this can occur at random with a probability of 0.0668 (or once out of 15 trials), which we would not regard as being significant.

4.1.2   Student's *t*

The approach outlined in the previous section is perfectly satisfactory if the number of items in each distribution being compared is "large", as the sample variances will approximate closely to the true population variances from which those samples were drawn; however, this is not satisfactory for "small" samples. This problem is overcome with the *t*-test, which was the invention of W. S. Gosset, a research chemist and not a mathematician, who very modestly published under the pseudonym of "Student" (1908). Student's *t* was later consolidated by Fisher (1925a,b), and it is similar to the standard error of difference method; it differs in that it takes into account the dependence of variance on sample size.

Student's *t* is defined formally as the absolute difference between means divided by the standard error of difference:

$$\text{Student's } t = \frac{|\bar{x}_1 - \bar{x}_2|\sqrt{N}}{\sigma},$$

where the expression $|\bar{x}_1 - \bar{x}_2|$ represents the absolute distance between the means (irrespective of sign) which is divided by the standard error of difference of the mean of the population, $\sigma/\sqrt{N}$ (this places $\sqrt{N}$ on the top of the fraction).

We now must be very clear about which population we are considering. When we use the *t*-test to distinguish between populations, we must assume the *null hypothesis*. This means that we start out believing there is no difference between the two populations. It's exactly the same principle as used in the judiciary of civilized societies, where an accused is assumed to be innocent until proven guilty. If, therefore, we assume there is no difference between the samples drawn from two supposedly different distributions or populations, there is no reason for not combining the two and then calculating a pooled variance. However, before we can do this we must further consider the dependence of variance on sample size.

We have seen earlier that standard deviation is dependent on the number of items in a sample (Section 3.1.2), hence variance is also dependent on sample size. The best estimate of the population variance, $\hat{\sigma}^2$, is given by

$$\hat{\sigma}^2 = \left(\frac{N}{N-1}\right) \text{Var}(s),$$

where the "hat", ^, over σ^2 means the best estimate, N is the number of items in the sample and $\text{Var}(s)$ is the variance of a sample obtained from the population. The expression $N/(N-1)$ is Bessel's correction (see Whittaker and Robinson 1960) which is particularly necessary whenever there is a small number of items in the sample. If the sample is large this expression tends to unity, and in such cases it is not necessary; however, it is better that it be included as this generalises the expression for samples of any size. We saw in Section 3.1.3 that the standard error for a sample of N items is given by the expression $s/\sqrt{N}$, hence the variance for N items, $\text{Var}(s)$, is given by the square of this, where $\text{Var}(s) = s^2/N$. If we now substitute this into the previous equation we get

$$\hat{\sigma}^2 = \left(\frac{N}{N-1}\right)\frac{s^2}{N} = \frac{s^2}{N-1},$$

where $\hat{\sigma}^2$ is the best estimate of the population variance obtained from s^2, which is the variance of the sample of N items drawn from that population.

We are now in a position to calculate the pooled variance of two samples which may or may not be drawn from two populations. The sums of the squared deviations from samples 1 and 2 are given respectively by the expressions

$$\Sigma(x_1-\bar{x}_1)^2 \quad \text{and} \quad \Sigma(x_2-\bar{x}_2)^2.$$

The total sample number, incorporating Bessel's correction for small sample sizes, is $(n_1-1)+(n_2-1)$ giving (n_1+n_2-2), which is referred to as the number of degrees of freedom. Thus, the best estimate $\hat{\sigma}^2$ for the pooled variance is given by

$$\hat{\sigma}^2 = \frac{\Sigma(x_1-\bar{x}_1)^2 + \Sigma(x_2-\bar{x}_2)^2}{n_1+n_2-2},$$

where $\bar{x}_1$ and $\bar{x}_2$ are the means of the two samples, and where x_1 and x_2 represent the individual deviations within the respective samples from their mean values. Now, as $s^2 = \Sigma(x-\bar{x})^2/n$ it follows that $\Sigma(x-\bar{x})^2 = n \times s^2$, hence the preceding equation may be rewritten as

$$\hat{\sigma}^2 = \frac{n_1 \times s_1^2 + n_2 \times s_2^2}{n_1 + n_2 - 2},$$

where s_1^2 and s_2^2 are the two individual sample variances. This best estimate $\hat{\sigma}^2$ of the pooled variance has been calculated from s_1^2 and s_2^2 with Bessel's correction for small sample numbers, but we must now relate $\hat{\sigma}^2$ to the total number of items in the combined sample to obtain the standard error of difference SE as follows:

$$SE = \sigma \sqrt{\frac{1}{n_1} + \frac{1}{n_2}}.$$

This expression for the standard error of difference for this particular problem is used in the calculation of Student's t, the complete formulation of which is

$$\text{Student's } t = |\bar{x}_1 - \bar{x}_2| \Big/ \sqrt{\left(\frac{n_1 \times s_1^2 + n_2 \times s_2^2}{n_1 + n_2 - 2}\right) \times \left(\frac{1}{n_1} + \frac{1}{n_2}\right)}.$$

The t-table, shown as a graph in Figure 4.2, is now entered with $(n_1 + n_2 - 2)$ degrees of freedom to obtain the probability that the calculated value of t could have arisen by chance.

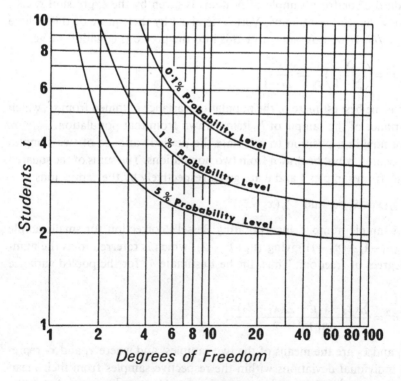

Figure 4.2. Student's t-values plotted against degrees of freedom, where probability levels at 0.05, 0.01 and 0.001 are shown as the curves.

As an example I will use the data in Section 3.1.2, where a sample of five items was generated randomly from a normal distribution containing 10,000 events with a mean value of 0.0 and standard deviation 200. The sample mean and standard deviation were 0.08 and 1.35 in standardized units, which respectively equate to 16 and 270. The difference in means was therefore 16, and we calculate the standard error of difference as

$$SE = \sqrt{\left(\frac{10,000 \times 200^2 + 5 \times 270^2}{10,000 + 5 - 2}\right) \times \left(\frac{1}{10,000} + \frac{1}{5}\right)} = 89.5.$$

Thus, $t = 16/89.5$ or 0.179, which is not even significant enough to get onto the t-table (see Figure 4.2); thus we can assume with a very high probability ($p > 0.9$) that the sample was in fact drawn from the parent population.

Let us now assume that the sample was made up of not five events but rather 1000. The standard error in this case will be

$$SE = \sqrt{\left(\frac{10,000 \times 200^2 + 1000 \times 270^2}{10,000 + 1000 - 2}\right) \times \left(\frac{1}{10,000} + \frac{1}{1000}\right)} = 6.87,$$

which gives $t = 16/6.87 = 2.32$. When we enter the t-table for this value with 10,998 degrees of freedom (i.e. very large) we find $p < 0.01$. Thus, this particular sample distribution would be expected to arise from the parent from which it was generated in less than 1 in 100 trials on a random basis, so we must conclude that there is a significant difference between the two. This may seem a little strange at first, because the sample was in fact generated from the parent normal distribution and thus should not be significantly different from that distribution. However, I'm sure you will appreciate that I've moved the goal posts in the middle of the game. If I had originally asked for a sample of 1000 events and not five then I would have obtained an entirely different sample distribution, and it is not legitimate arbitrarily to change the sample size and use parameters obtained with a different sample size.

4.1.3   Variance assessment

A tacit assumption in using the null hypothesis for Student's t is that there is no difference between the means. However, we also assumed that the variances of the two samples were not significantly different in calculating the pooled variance, and this must be shown to be true before using Student's t. There are two methods for doing this, the standard error of difference of the standard deviations of samples and the so-called F-test.

The standard-error-of-difference method uses exactly the same principles as described in the previous section, but with the correction for sample number described in Section 3.1.4. We will again take the data in Section 3.1.2, where a sample of five items was generated randomly from a normal distribution containing 10,000 events with a standard deviation of 200, yielding a sample standard deviation of 270. A pooled common variance Var is calculated as before, where

$$\text{Var} = \left(\frac{10,000 \times 200^2 + 5 \times 270^2}{10,000 + 5 - 2}\right) = 40,024,$$

which is the best estimate of the combined variance. We now correct for sample numbers to get the standard error of the difference of sample standard deviations, SE_s, where

$$\text{SE}_s = \sqrt{\text{Var}\left(\frac{1}{2n_1} + \frac{1}{2n_2}\right)} = \sqrt{40,024\left(\frac{1}{20,000} + \frac{1}{10}\right)} = 63.28.$$

The observed difference between the standard deviations is $270 - 200 = 70$, and dividing this by the standard error of difference of sample standard deviations, SE_s, gives 1.106. Thus, the observed difference is only 1.106 standard errors of difference and is not significant ($p > 0.05$).

The second method of assessing the variance involves the variance ratio test (Fisher and Yates 1963), which was designated the F-test by Snedcor (see Moroney 1990, p. 335) in recognition of the contribution made to statistical theory by Fisher. F is calculated as the ratio of the greater estimate of sample variance to the lesser estimate of sample variance. Again, we use the best estimate of sample variance, $\hat{\sigma}^2$, which is obtained using Bessel's correction as described in Section 4.1.2, where

$$\hat{\sigma}^2 = \left(\frac{N}{N-1}\right)\text{Var}(s).$$

In this example we calculate the greater variance estimate as

$$\frac{5}{4} \times 270^2 = 91,125$$

and the lesser variance estimate as

$$\frac{10,000}{9999} \times 200^2 = 40,004.$$

Thus, the variance ratio is $91,125/40,004 = 2.278$. It is worth noting that if we had failed to apply Bessel's correction we would have obtained $270^2/200^2 = 1.8225$, which is an underestimate. The relevant sections of the variance ratio tables are shown in Table 10 for ease of reference; more comprehensive tables are given in Appendix 3.

We now enter these tables with four degrees of freedom for the greater variance estimate and 9999 degrees of freedom (i.e. very large $= \infty$) for the lesser variance estimate, finding F values of 2.4 and 3.3 at the 5% and 1% probability levels, respectively. Thus, the ratio obtained, 2.278, does not reach the 5% probability level (i.e. $p > 0.05$), and hence this is again not significant.

4.2   Non-parametric

Non-parametric statistical tests are those developed to handle problems where the distribution from which a sample has been extracted is not known or

Table 10. *Variance ratio tables at the 5% and 1% probability levels for the degrees of freedom shown*

Degrees of freedom of lesser variance estimate	Degrees of freedom of greater variance estimate				
	1	2	3	4	5
5% probability level					
1	161	200	216	225	230
2	18.5	19	19.2	19.2	19.3
3	10.1	9.6	9.3	9.1	9.0
4	7.7	6.9	6.6	6.4	6.3
5	6.6	5.8	5.4	5.2	5.0
10	5.0	4.1	3.7	3.5	3.3
20	4.3	3.5	3.1	2.9	2.7
∞	3.8	3.0	2.6	2.4	2.2
1% probability level					
1	4052	4999	5403	5626	5764
2	98	99	99	99	99
3	34	31	29	29	28
4	21	18	17	16	16
5	16	13	12	11	11
10	10	7.6	6.6	6.0	5.6
20	8.1	5.8	4.9	4.4	4.1
∞	6.6	4.6	3.8	3.3	3.0

cannot be defined. In general these types of problems tend to be solved using ranking methods which analyse the order in which measurements, observations or value judgements occur. A good and a readily appreciated example is a panel of judges for something like that dreadful Eurovision song contest (this is a personal value judgement, not a statement of fact). The English judge will often place a female French contestant first, because she looks good and Englishmen like sexy female singers from France. If the contestant had been male he would probably have been placed last, as Englishmen don't like sexy male singers from France. The French judge will often place the English contestant last, usually quite correctly, because it's a load of rubbish. Very occasionally a majority of judges will place one particular contribution first, but no two judges will rank all the offerings in the same order. Clearly, this type of problem does not have a distribution that can be defined, but at the end of the day someone, for what it's worth, must be declared the winner. Non-parametric tests have been developed for these types of problems, although I suspect they are not rigorously applied in the Eurovision song contest.

4.2.1 Mann–Whitney

In the previous example all judges would give some score to all contestants, and the judgement consistency and relative agreement between judges

Table 11. *Two groups of*
numbers with five in one
group and four in the other

X-group		Y-group	
x_1	3	y_1	9
x_2	7	y_2	31
x_3	15	y_3	44
x_4	23	y_4	51
x_5	36		

Table 12. *Number groups from Table 11 ordered by rank*
positions

Y-group			y_1			y_2		y_3	y_4
X-group	x_1	x_2		x_3	x_4		x_5		
Values	3	7	9	15	23	31	36	44	51
Rank	1	2	3	4	5	6	7	8	9

can be determined with Spearman's rank correlation coefficient (see Section 4.2.3). However, we may be confronted with two sets of values from two possibly different groups which may or may not have been drawn from two different distributions, without any knowledge about the putative distributions from which those samples were drawn. This problem was originally addressed by Wilcoxon (1945) and was later refined by Mann and Whitney (1947).

Let us suppose we have two groups of numbers, conveniently christened X-group and Y-group, with five values in X and four values in Y as shown in Table 11. We now place these numbers in order according to magnitude, which for convenience is shown horizontally starting with the smallest number on the left (see Table 12). Also shown is the rank position of each number; for example, x_1 (the smallest number) is ranked 1, y_1 is ranked 3 as it is the third smallest and y_4, the largest number, is ranked 9.

We can see that the four values in the Y-group seem to be further towards the right compared with the five values in the X-group. The question is whether this particular arrangement could have occurred purely on a random basis, which is answered using the Mann–Whitney U-test. To do this we find out how many x-values lie to the right of every y-value and sum the result to obtain U for the Y-group, U_Y. There are three x-values (x_3, x_4 and x_5) to the right of y_1 and one x-value to the right of y_2. However, there are no x-values to the right of y_3 or y_4 as the latter are the two largest values in the rank, thus $U_Y = 4$.

If we have groups of small numbers such as these we can readily use this procedure to obtain U. But this is not convenient for sets of large numbers, and the following formula is used:

$$U_Y = N_X N_Y + \frac{N_Y(N_Y+1)}{2} - T_Y,$$

where N_X and N_Y are the number of values in the X- and Y-groups (respectively) and where T_Y is the sum of the rank positions for the Y-values. In this example $T_Y = 26$ which is the sum of $3+6+8+9$, the rank numbers of y_1, y_2, y_3 and y_4. Substituting the relevant parameters in the above relationship we obtain

$$U_Y = (5 \times 4 + 4(4+1)/2 - 26) = (20 + 10 - 26) = 4,$$

which is what we got by summing the total number of x-values lying to the right of every y-value.

We also need to carry out the same procedure for the X-group of numbers. There are four y values to the right of both x_1 and x_2 ($U_X = 8$ thus far), three y-values to the right of both x_3 and x_4 (U_X has now risen to 14) and two y-values to the right of x_5, which gives $U_X = 16$. The analytical formula for this value of U_X is given by

$$U_X = N_X N_Y + \frac{N_X(N_X+1)}{2} - T_X,$$

which is identical to the previous relationship for U_Y apart from substituting X's for Y's in the relevant places. T_X is the summation of the rank numbers of x_1, x_2, x_3, x_4 and x_5 ($1+2+4+5+7 = 19$), from which we obtain

$$U_X = (5 \times 4 + 5(5+1)/2 - 19) = (20 + 15 - 19) = 16,$$

which again is what we got before.

If the X- and Y-values are randomly distributed in the rank, the sum of the rank positions T has a mean value $\bar{T}$ and a variance σ_T^2, given by the following expressions:

$$\bar{T}_X = \frac{N_X(N_X+N_Y+1)}{2} \quad \text{or} \quad \bar{T}_Y = \frac{N_Y(N_X+N_Y+1)}{2}.$$

$\bar{T}$ will be different for the X- and Y-groups if the numbers in these groups are different, but the variance will be the same for each:

$$\sigma_T^2 = \frac{N_X N_Y(N_X+N_Y+1)}{12}.$$

If both samples are large (> 20) then we take the values of T and $\bar{T}$ associated with the smaller of the pair of U-values, which in this case would have been T_Y and $\bar{T}_Y$ if N_X and N_Y had both been greater than 20, in order to calculate the z-statistic which is given by

$$z = \frac{|T_Y - \bar{T}_Y|}{\sqrt{N_X N_Y (N_X + N_Y + 1)/12}}.$$

This expression should seem familiar, as we've seen something like it before in Student's t. The numerator $|T_Y - \bar{T}_Y|$ represents the difference between the value of T for the Y-group of values and the mean $\bar{T}$ that it would be predicted to have if the two groups of numbers were randomly distributed within the rank structure. The denominator is the square root of the variance, which is the standard deviation. Thus, z represents the observed deviation from the mean in standard deviation units. For large numbers z approximates very closely to the normal distribution; hence probability values can be associated with z by reading off from the cumulative frequency distribution of the normal curve.

This is not so for small sample numbers such as the data in Table 12. Mann and Whitney (1947) computed the various probabilities associated with U values for different sized samples. These are reproduced in full in Appendix 4, but Table 13 for $N_2 = 5$ (N_X in our previous notation) is shown here for ease of explanation. We now find that a U value of 4 is associated with $p = 0.095$ for n_1 (i.e. n_Y) equal to 4. Hence the distribution shown in Table 12 is not significant, as this would be expected to occur on a random basis once every nine trials.

An example of the use of the Mann–Whitney U-test for analysis of flow cytometric data is now given for p62$^{c\text{-}myc}$ quantitation in testicular tumours versus clinical outcome (Watson et al. 1986). The data are shown in Figure 4.3. The patients were divided into two groups, those who were alive and well (A/W) with no recurrence at 3 years after diagnosis and those who developed recur-

Table 13. *Mann–Whitney table for $N_2 = 5$*

U	n_1				
	1	2	3	4	5
0	.167	.047	.018	.008	.004
1	.323	.095	.036	.016	.008
2	.500	.190	.071	.032	.016
3	.667	.286	.125	.056	.023
4		.429	.196	.095	.048
5		.571	.286	.143	.075
6			.393	.206	.111
7			.500	.278	.155
8			.607	.365	.210
9				.452	.271
10				.548	.345
11					.421
12					.500
13					.579

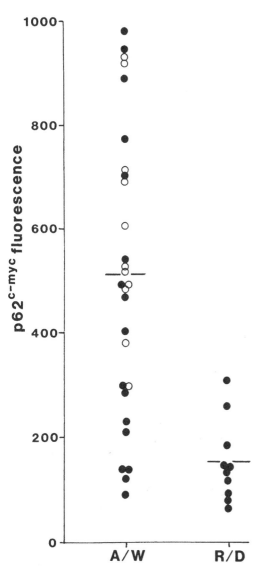

Figure 4.3. p62^{c-myc} levels for those patients who were alive and well (A/W) with no recurrence at 3 years after diagnosis and those who developed recurrence within this interval and subsequently died (R/D). The mean oncoprotein levels were 513 and 155 for the good and bad prognosis groups, respectively.

rence within this interval and subsequently died (R/D). The mean oncoprotein levels were 513 and 155 for the good and bad prognosis groups respectively. The Mann–Whitney test gave $U = 30$ with $n_1 = 10$ and $n_2 = 28$, and the large sample z was 3.65 ($p < 0.001$) that the observed distribution could have arisen on a random basis.

4.2.2 Kolmogorov–Smirnov

The Kolmogorov–Smirnov (K–S) statistic D represents a measure of the significance of the maximum vertical displacement between two cumulative frequency distributions. The one-tail test compares an experimentally derived with a theoretical cumulative frequency distribution, and the two-tail test compares two experimentally derived cumulative frequency distributions (see Chapter 6 in Siegel and Castellon 1988). In flow cytometry we should always be comparing a control with a test sample, so I will concentrate on the two-tail test. The K–S test was first used in flow cytometry by Young (1977), and it is particularly useful for comparing immunofluorescence control and test sample histograms.

The cumulative frequency distributions of the control containing n_1 cells and the test sample containing n_2 cells may be calculated as follows: for $1 \le i \le 256$,

$$F_{n_1}(i) = \sum_{j=1}^{j=i} f_{n_1}(j) \quad \text{and} \quad F_{n_2}(i) = \sum_{j=1}^{j=i} f_{n_2}(j).$$

This looks a bit like one of those awful equations that biologists have difficulty even looking at let alone understanding, so we'll take it one step at a time. We have a histogram with 256 channels, and each channel $f_n(i)$, where i has the range of 1 to 256, will have a number of cells in it. I should also draw your attention to the fact that it is important to distinguish between lowercase f and uppercase F in the following discussion. Just for convenience let us suppose that the first five channels have 1, 2, 3, 4 and 5 cells respectively, which are represented by the symbols $f_n(1)$, $f_n(2)$, $f_n(3)$, $f_n(4)$ and $f_n(5)$. (I've omitted the sub-subscripts 1 and 2 for clarity.) The summation always starts with $j = 1$ and increments until $j = i$. If we start off with $i = 1$ we sum only $f_n(1)$ into $F_n(1)$, hence $F_n(1) = f_n(1) = 1$. Next, i will be 2. Again, j starts at 1 which sums $f_n(1)$ into $F_n(2)$; then it increments to 2 which now sums $f_n(2)$ into $F_n(2)$, so the latter ends up with a value of $1 + 2 = 3$. When i is 3, we again start with $j = 1$ which initially makes $F_n(3) = f_n(1) = 1$; then we increment i to 2 and then to 3 which sums both $f_n(2)$ and $f_n(3)$ into $F_n(3)$, yielding a value of $1 + 2 + 3 = 6$. This process continues, and for $i = 5$ we sum $f_n(1)$, $f_n(2)$, $f_n(3)$, $f_n(4)$ and $f_n(5)$ (which have values of 1, 2, 3, 4 and 5 respectively) into $F_n(5)$ to yield 15 $(1 + 2 + 3 + 4 + 5)$. That equation really isn't so awful as it looks; it is just mathematical shorthand for this very simple process of successive addition:

$$i = 1, \quad F_n(1) = f_n(1)$$
$$i = 2, \quad F_n(2) = f_n(1) + f_n(2)$$
$$i = 3, \quad F_n(3) = f_n(1) + f_n(2) + f_n(3)$$
$$i = 4, \quad F_n(4) = f_n(1) + f_n(2) + f_n(3) + f_n(4)$$
$$i = 5, \quad F_n(5) = f_n(1) + f_n(2) + f_n(3) + f_n(4) + f_n(5).$$

Now that we have our cumulative frequency distribution, we normalize it so that $F_n(\max) = 1$, where "max" in this particular example is 5 but which would

Table 14. *Absolute and relative cumulative frequencies for the numbers* 1, 2, 3, 4 *and* 5

i	Cumulative frequency	Normalized cumulative frequency
1	1	0.067
2	3	0.200
3	6	0.400
4	10	0.667
5	15	1.000

be 256 in a flow cytometry histogram. This is carried out by dividing all the F_n's (i.e. $F_n(1), F_n(2), F_n(3), \ldots$) by $F_n(\max)$. This is shown in Table 14, where both the cumulative frequency and the normalized cumulative frequency distributions are given for the previous example.

When using K–S analysis we assume (yet again) the null hypothesis: the probability functions $P_1(j)$ and $P_2(j)$ which underlie the respective frequency density functions (the histograms) $f_{n_1}(j)$ and $f_{n_2}(j)$ are samples assumed to be drawn from the same population, so that

$$P_1(j) = P_2(j), \quad -\infty \le j \le +\infty.$$

The D-statistic is computed as the maximum absolute difference between the two normalized cumulative frequency distributions over the whole of the two distributions, where

$$D = \max_j |F_{n_1}(j) - F_{n_2}(j)|.$$

As with the Mann–Whitney U there is a variance, Var, associated with the assumed common population from which the two putative samples, containing n_1 and n_2 items respectively, are drawn. This is given by

$$\mathrm{Var} = \frac{(n_1 + n_2)}{n_1 \times n_2}.$$

The standard deviation s can now be found by taking the square root of this relationship; then dividing D by s gives D_{crit}, where

$$D_{\mathrm{crit}} = \frac{\max|F_{n_1} - F_{n_2}|}{\sqrt{(n_1 + n_2)/n_1 \times n_2}}.$$

This type of relationship, where we are dividing a difference by a measure of dispersion, has been seen in all the other statistical tests described previously. Two-tail critical values D_c for large sample numbers with their probabilities are

Table 15. *K–S critical values and confidence limits*

D_c	p
1.0727	0.200
1.2238	0.100
1.3581	0.050
1.5174	0.020
1.6276	0.010
1.7317	0.005
1.8585	0.002
1.9526	0.001

given in Table 15. If either n_1 or n_2 is less than 25 then the K–S tables shown in Appendix 5 must be used. However, this is unlikely to be necessary for flow cytometric data unless rare events are being studied.

Let us suppose that we wish to compare two typical immunofluorescence histograms each containing 10,000 cells, where the maximum difference D between the normalised cumulative frequency distribution is 0.021. The standard deviation s will be $s = \sqrt{20{,}000/(10{,}000 \times 10{,}000)} = 0.01414$. The ratio $D/s = 1.48$ which, as can be seen, lies between 1.3581 and 1.5174 which are associated with p equal to 0.05 and 0.02, respectively. Hence the null hypothesis that the sample distributions are drawn randomly from a single common distribution can be rejected at the 95% confidence interval but not at the 98% confidence interval, and we conclude that the two are different at $0.05 > p > 0.02$. If, however, there had been only 5000 cells in each sample then the standard deviation would have been 0.02 and the D/s ratio would have been 1.05 which does not reach D_c at $p = 0.20$; thus the null hypothesis would have been upheld, that is, the samples were very likely to have been drawn from a common population, $p > 0.20$.

4.2.3   Rank correlation

Correlations between two or more sets of judgements or measurements can be determined with Spearman's rank correlation coefficient (Kendall 1948). This enables an objective assessment to be made regarding two or more sets of value judgements (e.g., scores for contestants in ice skating, the Eurovision song contest, gymnastics, etc.) or regarding between-laboratory consistency in carrying out a particular investigation. We will take the last example as this is "closer to home", and develop the scenario where two laboratories are involved in a quality-control exercise. Each is given samples of the same set of 10 different specimens and is then asked to determine the proportion of cells exhibiting

Table 16. *Hypothetical results from two laboratories each measuring samples from the same 10 specimens*

	Sample									
	1	2	3	4	5	6	7	8	9	10
Lab A	.61	.23	.31	.11	.41	.19	.10	.03	.07	.17
Lab B	.54	.38	.42	.20	.36	.27	.21	.11	.14	.12

Table 17. *Rank differences and their squares for the data in Table 16*

	Sample									
	1	2	3	4	5	6	7	8	9	10
Lab A	10	7	8	4	9	6	3	1	2	5
Lab B	10	8	9	4	7	6	5	1	3	2
Rank difference, d	0	-1	-1	0	2	0	-2	0	-1	3
d^2	0	1	1	0	4	0	4	0	1	9

a specific surface immunophenotype. The "results", which are completely fictitious, are shown in Table 16.

When we look through these data we find that both laboratories score sample 8 with the lower proportions, 0.03 and 0.11 respectively, so in each case this would be ranked 1. Sample 9 from Lab A has the next lowest value (0.07) in that series and is ranked 2, but sample 10 from Lab B is ranked 2 in its series as it has the next lowest value (0.12) in the B set. A summary of all the ranking positions is shown in Table 17, together with the rank difference d obtained by subtracting the ranking of Lab B from Lab A, as well as the square of the rank differences. In terms of ranking alone, the two laboratories agree exactly for only 4 of the 10 specimens: 1, 4, 6 and 8. Spearman's rank correlation coefficient R is given by the expression

$$R = 1 - \left(\frac{6 \sum d^2}{n^3 - n} \right);$$

$\sum d^2$ is the sum of the squared rank differences and n is the total number of specimens. In this example $\sum d^2 = 20$ and $n = 10$, hence we have

$$R = 1 - \left(\frac{6 \times 20}{10^3 - 10} \right) = 1 - \left(\frac{120}{990} \right) = 0.8787.$$

Spearman's rank correlation coefficient was designed to have a value of +1 if two rankings are in complete agreement and −1 if those two rankings are in total disagreement. Hence, our result of 0.8787 would suggest that, in terms of ranking, the two laboratories are in reasonably close agreement. But we must ask if these distributions of ranking could have occurred on a random basis. As long as n is large (10 or more; that's why there were 10 samples!) we can use Student's t to assess the significance of R, where

$$\text{Student's } t = R \times \sqrt{\left(\frac{n-2}{1-R^2}\right)}.$$

If we now substitute 0.8787 for R and 10 for n in the previous equation, we obtain

$$t = 0.8787 \times \sqrt{\left(\frac{10-2}{1-0.772}\right)} = 5.2$$

with eight degrees of freedom. We now enter the t-table (Figure 4.2) with eight degrees of freedom and find that this value of t is associated with $p < 0.01$. Thus, a value of 0.8787 for R is highly significant, from which we can conclude that the rankings of the results from the two laboratories were in close agreement.

However, this does not tell us anything about the quality of the between-sample agreement from the two laboratories. We can do this with Student's t by analysing the differences between laboratories. Table 18 gives the difference between results by subtracting those of Lab B from those of Lab A together with the squared differences, d^2.

The mean difference $\bar{x}$ is calculated by summing the difference row and dividing by n, the number of samples, to give −0.052. If there are no differences between the laboratories then this mean value should not differ significantly from zero, since for random differences we would expect those differences to cancel. We now calculate the variance of the sample difference s^2 according to the convenient standard formula

$$s^2 = \frac{\sum x^2}{n} - \bar{x}^2,$$

Table 18. *Differences and their squares for the data in Table 16*

	Sample									
	1	2	3	4	5	6	7	8	9	10
Lab A	.61	.23	.31	.11	.41	.19	.10	.03	.07	.17
Lab B	.54	.38	.42	.20	.36	.27	.21	.11	.14	.12
Difference	.07	−.15	−.11	−.09	.05	−.08	−.11	−.08	−.07	.05
d^2	.0049	.0225	.0121	.0081	.0025	.0064	.0121	.0064	.0049	.0025

where $\sum x^2$ is equivalent to $\sum d^2 = 0.0824$ yielding

$$s^2 = (0.0824/10) - (0.052 \times 0.052) = 0.0055.$$

We now adjust this with Bessel's correction for the small sample number by multiplying by $n/(n-1) = (10/9) = 1.1111$ to get $0.0055 \times 1.1111 = 0.006111$; taking the square root gives $\hat{\sigma} = 0.0782$, where $\hat{\sigma}$ is the best estimate of the population standard deviation. Student's t can now be calculated as

$$\text{Student's } t = \frac{|\bar{x}_1 - \bar{x}_2|\sqrt{N}}{\hat{\sigma}},$$

and by substituting the various values we obtain

$$t = \frac{|0 - (-0.052)|\sqrt{10}}{0.0782} = 2.1$$

with nine degrees of freedom. This value of t with nine degrees of freedom does not quite reach significance at the 5% probability level, and we must conclude that the between-laboratory differences in values are not significant. This may seem a little difficult to accept when we just glance at the data in Table 15, as Lab B produced higher values than Lab A in 7 of the 10 determinations. However, this number of discrepancies with these magnitudes of differences could have occurred on a random basis in about 1 trial in 15. Hence we must conclude that, with these criteria, the two series were not significantly different in spite of any prejudice we may have. However, in a quality-control exercise like this we would be justified in setting more stringent criteria. If we now take a probability level of 0.1 for magnitude discrepancies between laboratories – which would not be unreasonable, as we know they should be getting the same answers – then we must conclude that there was something suspicious which needs investigating. This brings up the very important point that you can set your significance criteria at any level you wish. If you want very good between-laboratory consistency you would set your probability level at 0.95, meaning there would be a 95% chance that the two laboratories were the same.

4.3   Association, χ^2

Associations of two or more apparently related events are very frequent occurrences in everyday life. During my early teens I was often told to wear an overcoat when it was raining, as I would catch a cold if I got wet. It was interesting that these warnings were usually issued in the autumn and winter, occasionally in the spring but never in the summer. In the summer my brother, two cousins, Robbie Smith and I would frequently venture down to the local river with our pets, a dog, a cat and a duck. The duck and cat would go swimming but the dog was terrified of water and went in only if, very occasionally, he was pushed in. We five children never intended to go in the water, as the river came straight off the hillside and was always very cold, but we frequently all ended up in the "drink" after someone had pushed someone else in and a fight ensued.

There are a number of interesting features to this anecdote. First, why didn't I get any warnings to wear an overcoat in the summer when it was raining? It rained more often in the autumn, winter and early spring than in the summer, but it still did rain in the latter season. Second, why didn't I catch a cold in the summer, particularly as we would frequently get soaking wet in the river, while in contrast I would often catch a cold in the winter? Third, I did not catch a cold in the winter every time I went out in the rain without wearing an overcoat. Finally, why did the cat, quite voluntarily, go swimming? I'm not going to attempt to answer the last question; I can only assume that he liked it.

I recently found my first (and only) attempt at writing a diary, which reminded me of these events. It contained the usual sort of junk such as "got a bag of sweets today" which was a "happening" in the early 1950s when postwar rationing was being lifted. There were other entries such as "going back to school tomorrow, ugh"; "went down to river today, we all got wet, I got into trouble for pushing Keith and Angela in"; "got a cold, told to stay in bed"; "Jack Robinson let me drive the train to Haltwhistle"; "mucked out the pigs, very smelly, it was raining, got wet"; "helped with harvest, very hot"; etc., etc. Embedded within all this there were some data, albeit "low grade", concerning the frequency with which I caught a cold during that year, data that I was very roughly able to relate to my state of "wetness" or "dryness" as summarised in Table 19.

I recorded that I had caught a cold three times and that I had got wet a total of 20 times. I also recorded that each time I got a cold I had also got wet. The total number of entries in the diary was 51, but I did not record if I kept dry or if I didn't catch a cold. However, this is essentially a binomial problem, and I've assumed that I kept dry if I had not recorded that I got wet and that I had a cold only if this was recorded. Hence, all the frequencies in Table 19 can be determined. I'd now like to know if getting wet and catching a cold are associated, which is addressed by using χ^2.

χ^2 is used to test if the observed frequencies in a distribution differ from those to be expected on a random basis. The formula is

Table 19. *Observed frequencies*

	Caught a cold	Didn't catch a cold	Total
Got wet	3	17	20
Kept dry	0	31	31
Total	3	48	51

$$\chi^2 = \sum \frac{(O-E)^2}{E}.$$

The observed and expected frequencies are represented by O and E respectively, and I always remember that the expected frequency E should be on the bottom of the fraction because the observed frequency O could be zero and we are not allowed to divide by zero. The expected frequency will never be zero. Table 19 shows the observed frequencies, but how do we obtain those to be expected? If we take the first row, "got wet", we find that this happened 20 times. Thus, we would reasonably expect this frequency to be distributed between "caught a cold" and "didn't catch a cold" according to the ratio of the subtotal occurrence of these events to the total of all the events. "Caught a cold" occurred a total of 3 times out of a grand total of 51, and "didn't catch a cold" occurred 48 times out of 51. Thus, the total "got wet" frequency (20) must be distributed between "caught a cold" and "didn't catch a cold" in the ratios 3/51 and 48/51 respectively. These arguments also apply to the bottom row, where the 31 total occurrences of "kept dry" will also be distributed between "caught a cold" and "didn't catch a cold" according to the ratios 3/51 and 48/51.

A general schematic for calculating the expected frequencies is shown in Table 20, where there are eight columns $(A, \ldots, H)$ and four rows (W, X, Y, Z). The frequencies for the columns are given by $a, \ldots, h$ and for the rows by w, x, y, z. The total frequency for the whole table is Ω.

We now return to our specific problem and use this schematic to calculate the expected frequencies, which are shown in Table 21. We now have both the observed and expected frequencies, but first we must make a correction to the former because the χ^2 distribution is continuous whereas the frequencies in Table 19 are binomially distributed and thus are discontinuous. This involves adding 0.5 to those observed values that are below expectation and subtracting

Table 20. *General schematic for calculating expected χ^2 frequencies*

	A	B	C	D	E	F	G	H	Total
W	$\frac{aw}{\Omega}$	$\frac{bw}{\Omega}$	$\frac{cw}{\Omega}$	$\frac{dw}{\Omega}$	$\frac{ew}{\Omega}$	$\frac{fw}{\Omega}$	$\frac{gw}{\Omega}$	$\frac{hw}{\Omega}$	w
X	$\frac{ax}{\Omega}$	$\frac{bx}{\Omega}$	$\frac{cx}{\Omega}$	$\frac{dx}{\Omega}$	$\frac{ex}{\Omega}$	$\frac{fx}{\Omega}$	$\frac{gx}{\Omega}$	$\frac{hx}{\Omega}$	x
Y	$\frac{ay}{\Omega}$	$\frac{by}{\Omega}$	$\frac{cy}{\Omega}$	$\frac{dy}{\Omega}$	$\frac{ey}{\Omega}$	$\frac{fy}{\Omega}$	$\frac{gy}{\Omega}$	$\frac{hy}{\Omega}$	y
Z	$\frac{az}{\Omega}$	$\frac{bz}{\Omega}$	$\frac{cz}{\Omega}$	$\frac{dz}{\Omega}$	$\frac{ez}{\Omega}$	$\frac{fz}{\Omega}$	$\frac{gz}{\Omega}$	$\frac{hz}{\Omega}$	z
Total	a	b	c	d	e	f	g	h	Ω

Table 21. *Expected frequencies*

	Caught a cold	Didn't catch a cold	Total
Got wet	$\dfrac{3\times20}{51}=1.1765$	$\dfrac{48\times20}{51}=18.8235$	20
Kept dry	$\dfrac{3\times31}{51}=1.8235$	$\dfrac{48\times31}{51}=29.1765$	31
Total	3	48	51

Table 22. *Corrected observed frequencies*

	Caught a cold	Didn't catch a cold	Total
Got wet	2.5	17.5	20
Kept dry	0.5	30.5	31
Total	3	48	51

0.5 from those above expectation (Yates 1934); these corrected frequencies are shown in Table 22.

We can now calculate χ^2, which for the first "cell" is $(2.5-1.1765)^2/1.1765$, $(17.5-18.8235)^2/18.8235$ for the second, $(0.5-1.8235)^2/1.8235$ for the third, and finally $(30.5-29.1765)^2/29.1765$ for the last cell. This gives a total sum of χ^2, $\Sigma\chi^2$, of 2.603, and we now enter the χ^2 table shown as Figure 4.4 to find the probability level associated with this value. However, we also have to know the number of degrees of freedom, given by the product of the (number of rows -1) multiplied by the (number of columns -1), which for Table 20 would be $(4-1)\times(8-1)=21$. In our specific example (Table 19) the number of degrees of freedom is $(2-1)\times(2-1)=1$, and with this we find that 2.603 does not reach the 5% probability level. Hence, $p>0.05$ and the observed distribution in Table 19 could have occurred by chance in about one trial in nine. Thus, in this particular case, there was no significant association between getting wet and catching a cold, which I'm very relieved about because I never believed that story anyway.

The χ^2 statistic is very versatile and can be used not only for testing associations but also as a curve-fitting criterion (see Section 4.4.2). However, when we use this statistic to test the significance of associations we must always remember that a highly significant association can never be taken to imply a causal relationship. The sale of blankets in Montreal, before the days of modern heating

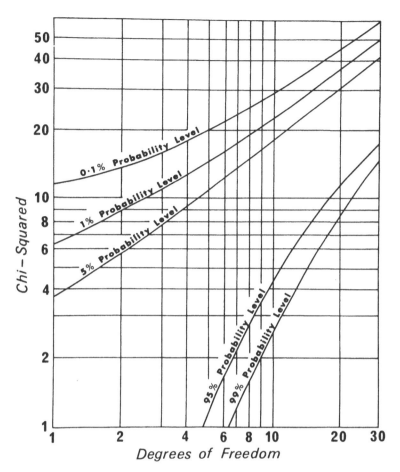

Figure 4.4. χ^2 (chi-squared) plotted against degrees of freedom, where probability levels at 0.99, 0.95, 0.05, 0.01 and 0.001 are shown as the curves.

and air conditioning, was correlated and very closely associated with the rainfall in Manchester, U.K. Nevertheless, nobody in their right mind would suggest either that people went out to buy blankets in Montreal *because* it was raining in Manchester or that it would start raining in Manchester *because* people were buying blankets in Montreal. It is just that there is a common denominator; both cities are at about the same latitude, and when the weather is awful in Manchester it is also awful in Montreal.

4.4 Fit criteria

A number of instances arise where we would need to fit a theoretical curve or distribution to a set of data points. Examples include fitting distributions to immunofluorescence data, fitting theoretical models to a single DNA

Table 23. *Six values for*
y associated with their
respective x values

x	y
0.5	0.50
1.0	0.65
1.5	2.50
2.0	4.50
2.5	5.50
3.0	9.25

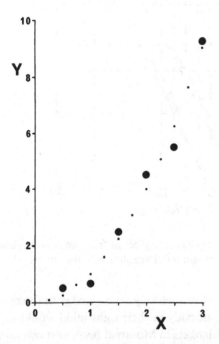

Figure 4.5. Data points from Table 23 which could be represented by the curve $y = x^{(k \times 2)} + c$. The small dots represent the curve with $k = 1$ and $c = 0$.

histogram in the process of deconvoluting the proportions in G1, S and G2 + M, and analysis of a time course of results from a series of DNA histograms to obtain cell-cycle phase times. Examples of all of these will be shown later, but first we consider the statistical criteria by which we fit the theoretical model to the experimental data.

4.4.1 Student's t

Consider the data shown in Table 23, which might represent experimentally derived values for y for given values of x. When we plot these data, as in Figure 4.5, we might come to the conclusion that the change in y could be represented by the following type of relationship:

$$y = x^{(k \times 2)} + c.$$

In fact, it is this relationship with $k = 1$ and $c = 0$ where a little random variation has been included. However, let us assume that we did not know this and proceed by guessing values for k and c, and that respective values of 0.75 and 0.25 were used to calculate values for y for the values of x shown in Table 23. We now compare these new y-values ("theoretical") with the "experimental" data shown in Table 23 by subtracting the theoretical values from the experimental to obtain the differences, and then square those differences; see Table 24.

I'm sure you will recognise that this is exactly the same procedure as described in the example in Section 4.2.3 comparing results from two laboratories. The variance of difference, s^2, is calculated with the formula

$$s^2 = \frac{\sum x^2}{n} - \bar{x}^2 = \frac{18.714}{6} - (1.039 \times 1.039),$$

where $\sum x^2$ is equivalent to $\sum d^2 = 18.714$, the mean difference $\bar{x} = 1.039$ and where $n = 6$, which gives the variance as 2.04. Again, we use Bessel's correction to give $2.04 \times 6/5 = 2.448$, and taking the square root gives $\hat{\sigma} = 1.565$. If there were no error then the mean of the difference would be zero, so we can now use Student's t where

$$t = \frac{|0 - 1.039|\sqrt{6}}{1.565} = 1.626.$$

Let us now make a second guess for the parameters of our theoretical curve and use $k = 1$ and $c = 0.25$; comparisons of these "theoretical" data with the "experimental" are shown in Table 25. We now have the mean difference $\bar{x} = -0.225$ and $\sum d^2 = 1.423$, hence $\hat{\sigma} = 0.473$ (after Bessel's correction). Thus $t = 1.165$, which is better than the first result. Hence, we would successively vary

Table 24. *Differences with their squares between experimental and theoretical values*

Experimental	0.500	0.650	2.500	4.500	5.500	9.250
Theoretical	0.604	1.250	2.087	3.078	4.203	5.446
Difference	−0.104	−0.600	0.413	1.422	1.297	3.804
d^2	0.011	0.360	0.170	2.021	1.683	14.469

Table 25. *Differences with their squares between experimental and theoretical values*[a]

Experimental	0.500	0.650	2.500	4.500	5.500	9.250
Theoretical	0.500	1.250	2.500	4.250	6.500	9.250
Difference	0.000	−0.600	0.000	0.250	−1.000	0.000
d^2	0.000	0.360	0.000	0.063	1.000	0.000

[a] Theoretical values were calculated with a set of constants different from those used in Table 24.

k and c until the minimum value of t was found, as this would represent the most significant fit between theoretical and experimental (see Section 5.2.2).

4.4.2 $\Sigma \chi^2$

Just as the sum of squared errors, and hence Student's t, can be used to assess the goodness of fit in a curve-fitting procedure, so too can $\Sigma \chi^2$. I will re-use the data from the previous section where the words "experimental" and "theoretical" are replaced (respectively) by "observed" and "expected", as shown in Table 26. The top three rows are exactly the same as in Table 24 but the bottom row, instead of showing the squared difference, now shows χ^2. The latter is calculated very simply by dividing the squared differences by the expected values. We now sum the individual values of χ^2 to obtain $\Sigma \chi^2$, which is 4.091 with $n-1=5$ degrees of freedom. Referring to Figure 4.4, we can see that this is associated with a p value of about 0.5.

Table 26. χ^2 *for the differences between the sets of values shown in Table 24*

Experimental	0.500	0.650	2.500	4.500	5.500	9.250
Theoretical	0.604	1.250	2.087	3.078	4.203	5.446
Difference	−0.104	−0.600	0.413	1.422	1.297	3.804
$d^2/E = \chi^2$	0.0167	0.2880	0.0804	0.6547	0.4024	2.649

Table 27. χ^2 *for the differences between the sets of values shown in Table 25*

Experimental	0.500	0.650	2.500	4.500	5.500	9.250
Theoretical	0.500	1.250	2.500	4.250	6.500	9.250
Difference	0.000	−0.600	0.000	0.250	−1.000	0.000
$d^2/E = \chi^2$	0.0000	0.5538	0.0000	0.0139	0.1818	0.0000

This will now be repeated for the data in Table 25, where again the first three rows shown in Table 27 are the same as in Table 25, but where the last row is now χ^2 obtained by dividing the squared difference d^2 by the expected value. Summing the individual χ^2 values we get $\Sigma \chi^2 = 0.7495$ which is associated with $p > 0.95$, which is considerably better than before. Thus, just as with Student's t in the previous section, k and c are varied to obtain the values which minimize $\Sigma \chi^2$. This exercise will be considered further in Section 5.2.2.

5

Regression analysis

Regression analysis is used to establish a mathematical relationship between two variables. The easiest example to understand is the cost of purchasing. If we go into the retail outlet mentioned in Section 2.1 to buy an undergarment (you thought I'd forgotten) it may well cost $6.00. If we buy two they will cost $12.00 and three will cost $18.00. There is a direct functional relationship here between total cost and number of items purchased, where there is an increment of $6.00 for each item purchased. This process can be represented by

$$y = mx + c,$$

which is the formula for a straight line. The cost per item is the slope m, the number of items we purchase is x, c is the intercept on the Y-axis (which in this case is zero) and y is the total cost of x items. This process is represented in Figure 5.1. Not all transactions are this straightforward. If I buy one bag containing 10 lbs of potatoes at our local greengrocer it will cost me $0.90. If I buy six bags they cost me $0.85 each and eleven bags or more will cost $0.80 each. However, if I go round to the local farmer and buy 250 lbs in a large sack (with a certain amount of soil and some worms still attached) it costs me $12.50. Let's graph this result in Figure 5.2.

In the first example (Figure 5.1) there is a direct relation which is clearly apparent. There are also linear relationships over the first three segments in the second example (Figure 5.2). However, if this had been a set of scientific measurements (Y-axis) for various X-axis "inputs" then we would probably have interpreted this as an overall linear relationship or possibly some sort of curve and ignored the small discontinuities at X-axis scale readings of 50 and 100, explaining these away as "experimental variation". We might also have been tempted to interpret the 250-lb point as an "outlier" and ignored it, as "obviously!" it does not fit into the overall pattern. It is only by knowing the potato purchasing rules of the greengrocer and farmer that we are able to interpret fully the data in Figure 5.2. The problem with most scientific investigations is that we don't know all rules of the system we are studying – which is why we

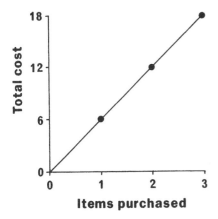

Figure 5.1. y (total cost) as a direct function of x (items purchased).

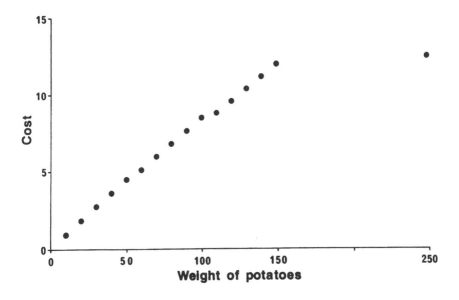

Figure 5.2. Cost (y-axis) of potatoes with discounts for buying in bulk (x-axis).

are doing the experiments – and in attempting to find the rules we must interpret data such as those in Figure 5.2.

Regression analysis is the means by which we can find the optimum relationship between a set of numbers (data) and an independent variable, where the data appear to vary in proportion to the independent variable. The relationship is always some form of mathematical function, such as the formula for a straight line $y = mx + c$ or the curve $y = x^{(k \times 2)} + c$ mentioned in Sections 4.4.1 and 4.4.2. However, we must be very clear that, even if we can obtain a very good fit and correlation between the mathematical relationship and the data, this does not mean that this is the "true" way that the observations behave in relation to the independent variable.

5.1   Linear regression

The points in Figure 5.3 are clearly related, in the sense that as x increases so does y. Furthermore, it would seem that there is a linear relationship with no "by eye" suspicion that there could be a curve. This (and similar) data sets would be most conveniently summarized if we had a mathematical function to describe the increase in y with the increase in x. If we now assume that this putative relationship is linear, we can proceed by using the formula for the straight line, $y = mx + c$, as the summary for the data. If, subsequently, we need to refer to the data then we could use the summary as opposed to *all* the data points, since the regression line is defined completely by $\bar{x}$, $\bar{y}$ (the means of all the x and y points, respectively), the slope m, and the intercept c on the Y-axis. We use linear regression analysis for this, where the squared deviations of the points from the line are minimised.

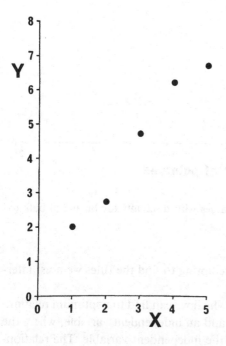

Figure 5.3. Various y-values plotted against their corresponding x-values.

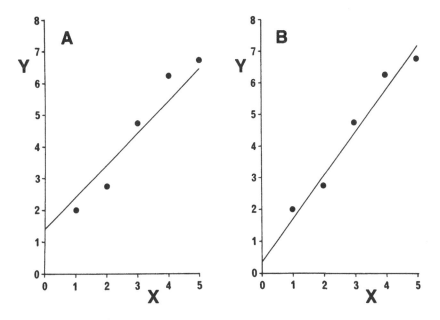

Figure 5.4. Panel A, a bad straight-line fit to the data in Figure 5.3; panel B, the regression analysis.

5.1.1 Minimising squared deviations

The principle of minimising squared deviations, least squares analysis, was introduced by Legendre (1806). In order to illustrate the processes involved, I've drawn two lines through the data points of Figure 5.3; these are shown in panels A and B of Figure 5.4. On inspection you would say, without any hesitation, that the fit in panel B is better than in A. If I were now to ask why, you might find it a little difficult to give an exact answer and reply non-specifically with a phrase such as "well I can see that it's better"; or "it's obvious"; or possibly "there are more points nearer to the line in B than in A". In fact, the human eye and mind are really quite good at drawing a best-fit line through data points such as these, and we've all done it using a transparent plastic ruler. It's much more difficult with a non-transparent ruler, as we then cannot see all the data points and so cannot adjust the position of the ruler's edge to make it as nearly adjacent to as many points as possible. The mathematical processes involved in regression analysis do exactly this by making the sum of squared differences of the points from the line as small as possible. In comparison, the eye and mind tend to make the absolute distances between points and line as small as possible.

The absolute Y-axis distances between the lines and the data points have been drawn as the vertical lines in the two panels of Figure 5.5. When we carry out a

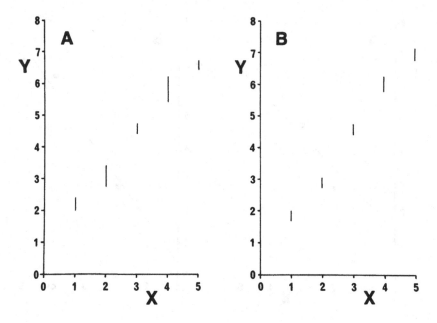

Figure 5.5. The absolute vertical distances between the lines in Figure 5.4 and the data points.

"regression analysis" by eye, we make a mental summation of all these absolute distances and come to the conclusion, quite correctly, that the summed deviations in panel B are smaller than in panel A and that the latter is the better fit. This, however, pays equal weight to all the deviations, whereas the mathematical process – minimising the sum of squared deviations – places relatively more weight on larger deviations. This is a process with which the human mind has difficulty, as we are not familiar with thinking in terms of squared distances.

We could proceed to find the minimum squared deviation by drawing multiple lines, finding the absolute vertical displacements, squaring these, summing them and continuing until we found the minimum squared deviation. Obviously, this would be very time-consuming, and we would find it difficult to be completely sure we had found the absolute minimum squared deviation. Needless to say, the mathematicians have been at this problem and have produced formulae to find m and c which give the minimum squared deviation. I'm not going through the derivations of these; they are in the big math tomes and are pretty heavy going, so I'll just state the formulae:

$$m = \frac{\sum(xy) - (\sum x) \times (\sum y)/N}{\sum(x^2) - (\sum x)^2/N};$$

$$c = \frac{(\sum x) \times \sum(xy) - (\sum y) \times \sum(x^2)}{(\sum x)^2 - N \times \sum(x^2)}.$$

It is necessary to be very clear about the meanings of each of these terms:

N = the number of pairs of points;
$\sum x$ = the sum of the x-values;
$\sum y$ = the sum of the y-values;
$\sum(xy)$ = the sum of the $x \times y$ products of all the pairs of values;
$(\sum x)^2$ = the square of the sum of the x-values;
$\sum(x^2)$ = the sum of the squares of the x-values.

$\sum(xy)$ is calculated by multiplying each (x, y)-pair of values and summing the N results. $(\sum x)^2$ is calculated by squaring the sum of all the x-values, and $\sum(x^2)$ is calculated by squaring each of the x-values then summing the N results.

5.1.2 Correlation coefficient

The regression line of panel B in Figure 5.4 was calculated with the preceding formulae, and we must now ask how good the correlation is and what confidence limits are associated with the estimates obtainable from the regression. This is carried out using the correlation coefficient r, which has been designed to have a value of unity if there is perfect correlation and of zero if there is no correlation. Negative values occur when the slope m is negative, and -1 would indicate perfect correlation with an inverse relationship between y and x. The formula for r is

$$r = \frac{1}{N} \sum\left(\left(\frac{x - \bar{x}}{\sigma_x}\right) \times \left(\frac{y - \bar{y}}{\sigma_y}\right)\right),$$

where $\bar{x}$ and $\bar{y}$ are the respective means of all the x and y values and where σ_x and σ_y are the respective standard deviations of the x- and y-values. The value of r for the regression in Figure 5.4B was 0.968 which indicates, as we can see, a very good correlation.

The functional relationship depicted by a regression line for the behaviour of y in relation to x will never be perfect in the real world, but it gives some indication of that behaviour. Moreover, if we wish to make estimates of y-values from the regression line for given x-values, we need to know by how much those estimates might be in error. This is addressed using the standard error of estimate S_y, which is given by

$$S_y = \sigma_y\sqrt{(1 - r^2)}.$$

Figure 5.6 displays panel B of Figure 5.4 with the addition of two dashed lines. These have been drawn parallel to the regression line at distances of ± 2 standard errors of the estimate (calculated from the equation for S_y) from the regression line. By referring to Appendix 2 we find that $z = 2$ (i.e. two standard errors) is associated with 0.0228. As the limits in Figure 5.6 have been drawn at $\pm 2 \times S_y$, we can calculate a probability of $1.0 - (0.028 \times 2) = 0.9544$ (i.e. 95%) that any

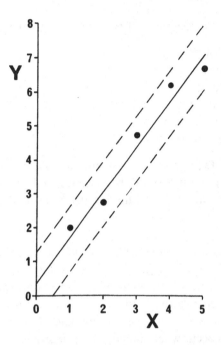

Figure 5.6. Panel B of Figure 5.5, with the addition of two dashed lines drawn parallel to the regression line at distances of ±2 standard errors of estimate.

estimate of y from the regression line for a given value for x will be within these limits.

5.1.3   Significance of the correlation coefficient

Apart from wanting to know the confidence limits for estimates of y for various x-values, we also wish to know the significance of the correlation coefficient. Clearly, if r has a low significance value then we will not be able to be very confident about the estimates of y from x. The standard error of estimate of r is given by the expression $\sqrt{(1-r^2)}$, hence we can calculate Student's t for r as

$$t = \frac{r \times \sqrt{(N-2)}}{\sqrt{(1-r^2)}},$$

where $N-2$ is the number of degrees of freedom. By substituting the values for N and r (15 and 0.968) in this relationship we get a value of $t = 11.57$ for the data in Figure 5.4B, and entering the t-table (Figure 4.1) with three degrees of freedom (five points) we find $p < 0.01$.

5.1.4   Regression of x on y

In the previous three sections we have covered regression of y on x, albeit somewhat superficially. This is not the same as x on y, where we would be estimating x-values for given y-values from the regression. In the y-on-x regression we minimised the squared deviations in the vertical (Y-axis) direction; in x-on-y regression we have to minimise the squared deviations in the horizontal (X-axis) direction. These are identical if and only if the correlation coefficient is unity – meaning there is a perfect functional relationship between x and y. I'm not going to write out the whole of the text again for the x-on-y regression merely because of a 90° "phase shift", so I'll just list the relevant equations:

$$m = \frac{\Sigma(xy) - (\Sigma x) \times (\Sigma y)/N}{\Sigma(y^2) - (\Sigma y)^2/N};$$

$$c = \frac{(\Sigma x) \times \Sigma(xy) - (\Sigma x) \times \Sigma(y^2)}{(\Sigma y)^2 - N \times \Sigma(y^2)};$$

$$r = \frac{1}{N} \Sigma\left(\left(\frac{x - \bar{x}}{\sigma_x}\right) \times \left(\frac{y - \bar{y}}{\sigma_y}\right)\right);$$

$$S_x = \sigma_x \sqrt{(1 - r^2)};$$

$$t = \frac{r \times \sqrt{(N-2)}}{\sqrt{(1 - r^2)}}.$$

A FORTRAN routine for calculating a linear regression of y on x with the correlation coefficient and standard error of estimate is given in Appendix 6.

5.1.5 Testing regression linearity

At the outset of this Section 5.1 we *assumed* that the data could be described by the equation for a straight line, $y = mx + c$, but this assumption has not yet been shown to be compatible with the data. However, in some data sets this can be tested for with a parameter called the *correlation ratio,* which is used to compare the variance due to deviations of the means of arrays from the regression line with the variance due to deviations within arrays. This last statement is a bit of a brainfull so let us look at the meaning in more simple language.

Consider Figure 5.7, where there are four y-values for each x-value. Each of the groups of y-values has a mean represented by a horizontal line. This is a regression line calculated from all the data points drawn through the whole lot. None of the means of the three arrays of y-values lies on the regression line. Hence, there is a dispersion of these mean values about the regression line. This dispersion can be associated with a variance which represents "the variance due to deviations of means of arrays from the regression line", as stated in the first component of the description of the linearity test that uses the correlation ratio. There is also dispersion within each of the arrays, which is the second component in the description of the linearity test – namely, "variance due to deviations within arrays". To cut a long story short (the long story is in Appendix 7 if you really want it), we compare the two variances with the F-ratio, where

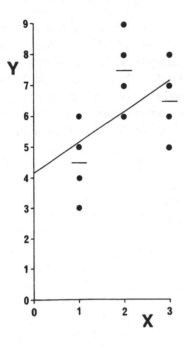

Figure 5.7. Three groups of y-values each with four values at each x-value.

$$F = \frac{\eta^2 - r^2}{1 - \eta^2} \times \frac{N - I}{I - 2}.$$

We have previously met r, the correlation coefficient, but there is a new symbol here, η, the correlation ratio which will be defined later. The two expressions $I - 2$ and $N - I$ are the numbers of degrees of freedom in greater and lesser variance estimates (respectively), but what exactly are I and N? N is the total number of points, and the total number of degrees of freedom is $N - 1$. One degree of freedom is associated with the linear regression, which now leaves $N - 2$ degrees of freedom for the within-array variation plus the deviations of means from the regression line. There are I arrays of y's, hence we have $N - I$ degrees of freedom associated with the variation within arrays and $I - 2$ degrees of freedom associated with the deviations of means from the regression line.

The correlation ratio η is given by the ratio σ_w / σ_y, where σ_w is the standard deviation of the weighted means of the arrays of y's and where σ_y is the standard deviation of all the y-values. A value for σ_w^2 can be obtained from

$$\sigma_w^2 = \left(\frac{1}{N} \sum_1^I \frac{A_i^2}{n_i} \right) - \bar{y}^2.$$

In this expression N is the total number of points, n_i is the number of points in the ith array of y's, A_i is the sum of the J y-values in the ith array of y's and $\bar{y}$ is the mean of all the y-values.

Table 28. *Illustration of the calculation for the sum of squares of the J y-values in the i th array of y's*

i	x	The J y-values at each x_i-value	A_i	A_i^2	$n_i = 4$ A_i^2/n_i
1	1	3, 4, 5, 6	$3+4+5+6=18$	324	81
2	2	6, 7, 8, 9	$6+7+8+9=30$	900	225
3	3	5, 6, 7, 8	$5+6+7+8=26$	676	169
			Sum of the three (A_i^2/n_i)-values $=475$		

In the example shown in Figure 5.7 there are three arrays of y's each of which contains J values, and in each case $J = 4$. Hence, A_i is the sum of the four y-values in each of these arrays. Each of these three A_i-values would then be squared and divided by n_i, the number of values in each set, which in this example is four for each. The summation is then effected by adding each of these three results. The y-values corresponding to the three x-values in Figure 5.7 are shown in Table 28 together with the A_i summations, A_i^2, A_i^2/n_i and the final summation of the three A_i^2/n_i-values.

This method for assessing the validity of the assumption of linearity can be used only if there are more y-values than x-values. Obviously, this can occur only if there are replicate y-values associated with some of the x-values, as would occur in replicate experiments. An example of the use of this method will be given in Section 5.2.1.

5.2   Non-linear regression

Non-linear regression analysis is essentially identical to linear analysis, where the sum of squared deviation is minimised. However, many non-linear equations can be transformed to linear functions with a little bit of ingenuity and mathematical manipulation, and linear regression analysis can then be performed. However, a word of warning must be sounded about statistical assessment of post-transformation data. The statistics covered in the previous sections of this chapter pertain to dispersion of data about a mean, where the deviations are compatible with a normal distribution. If, therefore, there are a number of different Y-axis data points for each X-axis point, as would be obtained in replicate experiments, and the data are compatible with a normal distribution before transformation, they may well be incompatible after transformation. Hence, when dealing with such data sets it is necessary to test for compatibility with a normal distribution both before and after the transformation. This can be carried out using a normal distribution random-number generator to simulate a distribution of n values with a mean corresponding to that of the data points and then comparing variances, as detailed in Section 4.1.3.

5.2.1 Data transforms

We will consider four mathematical functions of varying complexity which are encountered within the general area of cancer biology and radiation research. Some of these are directly related to analysis of flow cytometry results, including first-order kinetic loss of a fluorochrome with time and fluorescence increasing exponentially to an asymptote. One of the functions, the multitarget cell-survival curve, is not directly related to flow cytometry but was included because there are many people using these instruments who are also doing things like cell-survival work and so it might prove useful to someone.

Inverse exponential

This first function might represent first-order cell killing due to some toxic agent, or first-order loss of fluorochrome across the external cell membrane. It has the following formula:

$$y = e^{-(k \times t)}.$$

The linear transform of this equation is very simple, familiar to us all, and is effected just by taking logs to the base e to give

$$\log_e(y) = -(k \times t).$$

Thus, we take the $\log_e$ of the Y-axis points and then perform linear regression analysis on these transformed data. The resulting slope will be $-k$, and the intercept should not be significantly different from zero if the biology is behaving in the way that the mathematical relationship predicts it should. This can be tested for by using the standard error of estimate, SE_y, given in Section 5.1.2. The intercept c on the Y-axis which is obtained from the regression is compared with SE_y to obtain the number of standard errors of estimate by which the intercept differs from zero. The probability associated with this difference can now be obtained from the table of z in Appendix 2.

Inverted exponential

The second function describes an inverse exponential increase to an asymptote, which has the form

$$y = Y_0 \times (1 - e^{-kt}),$$

where Y_0 is the asymptote. A little more manipulation is required to convert this to linear form. We first divide both sides by Y_0 to give

$$\frac{y}{Y_0} = 1 - e^{-kt},$$

and on rearrangement we get

$$1 - \frac{y}{Y_0} = e^{-kt}.$$

The linear transformation is now completed by taking logs to the base e of both sides, yielding

$$\log_e\left(1 - \frac{y}{Y_0}\right) = -(k \times t).$$

The position here is a little different from the first example, as the transform on the left-hand side depends on the value of the asymptote Y_0 and the regression will pass through the origin ($c = 0.0$) for a unique value of this parameter. We proceed by guessing a value for Y_0 which more than likely will give a non-zero value for c. If this intercept is positive then the value of Y_0 must be increased to reduce the value of c; if the intercept is negative then Y_0 must be reduced. An iterative process is used to implement this with the increment for Y_0 maintaining its magnitude until a change in sign of c is encountered, at which point the increment of Y_0 is halved. Hence, by successive approximation the value of c approaches zero and continues until c reaches an arbitrarily determined small deviation from zero. This is illustrated in Table 29, where the data from the potato problem are used. You will remember that there were three linear segments

Table 29. *Results of the iteration for fitting the inverted exponential to the potato data*

Slope	Intercept	R	Y_0	Y_0 increment
$-4.560E-03$	$2.206E-02$	-0.9978	2400	2400.000
$-1.885E-03$	$-1.269E-03$	-0.9994	4800	1200.000
$-2.663E-03$	$2.735E-03$	-0.9992	3600	600.000
$-2.207E-03$	$9.155E-05$	-0.9993	4200	300.000
$-2.033E-03$	$-6.958E-04$	-0.9994	4500	300.000
$-2.207E-03$	$9.155E-05$	-0.9993	4200	150.000
$-2.116E-03$	$-3.338E-04$	-0.9994	4350	75.000
$-2.161E-03$	$-1.298E-04$	-0.9993	4275	37.500
$-2.183E-03$	$-2.143E-05$	-0.9993	4238	37.500
$-2.207E-03$	$9.155E-05$	-0.9993	4200	37.500
$-2.183E-03$	$-2.143E-05$	-0.9993	4238	18.750
$-2.195E-03$	$3.441E-05$	-0.9993	4219	9.375
$-2.189E-03$	$6.318E-06$	-0.9993	4228	4.688
$-2.186E-03$	$-7.600E-06$	-0.9993	4233	4.688
$-2.189E-03$	$6.318E-06$	-0.9993	4228	2.344
$-2.188E-03$	$-6.855E-07$	-0.9993	4230	1.172

in this problem, but treating it as a set of scientific observations (without knowing the potato selling rules) we would be justified in assuming this might be represented by a curve and that I just happen to have chosen this particular curve to fit to the data.

The iteration was started with both Y_0 and its increment equal to 2400, double the highest y-value. In only 16 iterations the procedure converged to an intercept of -6.855×10^{-7}, the absolute value of which is less than the arbitrary limit that was set "small" at 10^{-6}. This intercept was associated with a slope of -2.188×10^{-3}, a correlation coefficient r of -0.9993, an asymptote Y_0 of 4230 and a final increment of 1.172. We now use these values for Y_0 and the slope m to generate the theoretical best-fit Y-values, Y_{expct}, at each of the X-values; these are compared with the observed values, Y_{obs}, in Table 30. Also shown are columns of difference between the theoretical and observed values and the squared difference, d^2.

We can now see that the differences, when compared with the magnitudes of the observed values, are essentially trivial. Moreover, when we sum the difference column we get a total difference of -0.43 and hence an average difference of -0.0287, which shows that the regression line effectively passes through the middle of the data points with any under and over estimations tending to cancel out. Furthermore, we can calculate the variance of the difference, s^2, as in Section 4.2.3, with the formula

Table 30. *Differences between the best-fit theoretical curve and the potato data*

X	Y_{obs}	Y_{expct}	Difference	d^2
10	90	91.5	−1.50	2.25
20	180	181.1	−1.10	1.21
30	270	268.7	1.67	2.79
40	360	354.5	5.50	30.25
50	450	438.3	11.70	136.89
60	510	520.4	−10.40	108.16
70	595	600.7	−5.70	32.49
80	680	679.2	0.80	0.64
90	765	756.1	8.90	79.21
100	850	831.3	18.70	349.69
110	880	904.8	−24.80	615.04
120	960	976.8	−16.80	282.24
130	1040	1047.4	−7.40	54.76
140	1120	1116.7	3.30	10.89
150	1200	1183.3	16.70	278.89
Total			−0.43	1985.40

$$s^2 = \frac{\sum x^2}{n} - \bar{x}^2,$$

where $\sum x^2$ is equivalent to $\sum d^2$ and in Table 27 this is 1985.4. This gives $s^2 =$ (1985.4/15) − (−0.0287 × −0.0287) = 132.36, and taking the square root after Bessel's correction for small numbers gives $\hat{\sigma} = 11.91$. Using this for Student's t, we get $t = (0.0287 \times \sqrt{15})/11.9 = 0.0093$. This is a highly significant fit, $p >$ 0.99, so by all criteria the fit is good in spite of the fact we have used a mathematical function known to have absolutely no correspondence with the true relationship between x and y. Moreover, when a linear regression was carried out with the non-transformed data, $\sum d^2$ was 17,015 compared with 1985 for the curve used. It just so happens that we can find a particular segment of the curve used for the analysis which fits the data very well. It is also worth noting that the biggest discrepancies between the data and the regression curve occur in the vicinity of the discontinuities.

Multitarget cell-survival curve

The third linear transform to be demonstrated is that for the multitarget radiation cell-survival curve (Watson 1978). This function has a "shoulder" with an increasing slope that tends to exponential decline at high values on the X-axis (radiation dose). The exponential section of the curve extrapolates to a value greater than 1 on the ordinate, which is called the extrapolation number N. The equation is

$$s = 1 - (1 - e^{-D/D_0})^N.$$

This is a little more difficult than the last example but the difficulty is only quantitative, not qualitative. We start by rearranging the equation to give

$$1 - s = (1 - e^{-D/D_0})^N.$$

Next we take logs to the base e to get

$$\log_e(1-s) = N \times \log_e(1 - e^{-D/D_0}).$$

If we now divide through by N then exponentiate both sides, we obtain

$$\exp\left(\frac{\log_e(1-s)}{N}\right) = 1 - e^{-D/D_0}.$$

This is now rearranged to give

$$1 - \exp\left(\frac{\log_e(1-s)}{N}\right) = e^{-D/D_0}.$$

The linear transformation is now completed by taking the $\log_e$ of both sides:

$$\log_e\left(1 - \exp\left(\frac{\log_e(1-s)}{N}\right)\right) = \frac{-D}{D_0}.$$

There are two unknown quantities (N and D_0) in this equation, which is of the form $y = mx + c$. The left-hand side is equivalent to y and the transformed data are dependent on the value of N. The slope m is given by $-(1/D_0)$, and the intercept c will be zero for a unique value of N.

An initial guess for N can be made which, after the linear transform and regression analysis, is likely to be associated with a non-zero value of c. This, however, will give us values for m and $\bar{x}$ which, in association with the non-zero value for c, can give a better approximation for the extrapolation number N_1 from the following relationship:

$$N_1 \times \log_e(1 - \exp(\bar{x} \times m)) = N \times \log_e(1 - \exp((\bar{x} \times m) + c)),$$

where $\bar{x}$ is the mean of all the x-values, m is the initial approximation for the slope and N is the initial approximation for the extrapolation number that gave rise to the non-zero value for c. This better approximation for N is then used to re-transform the data, and this process continues until c is zero.

This example was included to demonstrate that even fairly complex equations can be transformed to linearity with a little cunning; its use, or in this particular case its misuse, is shown in Figure 5.8. The data used were obtained from EMT6/M/CC cells grown in tissue culture and irradiated with varying doses of 250-KV X-rays, and the figure shows the surviving-fraction transforms plotted against X-ray dose. The regression line passed through the origin with an extrapolation number N of 15.7 (5–50, 95% confidence limits) and an exponential slope D_0 of 149 (123–189, 95% confidence limits). The correlation coefficient r was -0.9795, indicating a very high degree of correlation. Moreover, Student's t (calculated as described in Section 5.1.3) was 167 with 50 degrees of freedom, which is associated with $p < 0.00001$ that this degree of correlation could have arisen by chance.

On the face of it, we might feel content with this result with such a high degree of statistically significant correlation. However, just looking at the data in Figure 5.8 should raise some doubts about the validity of applying a linear regression analysis, as the transformed data points do tend to describe a curve. A further analysis was carried out to test for linearity using the method described in Section 5.1.5. The correlation ratio was 0.9897 with 9 degrees of freedom in the greater and 41 degrees of freedom in the lesser variance estimate, hence

$$F = \frac{\eta^2 - r^2}{1 - \eta^2} \times \frac{N - I}{I - 2} = \frac{(0.9897^2 - (-0.9795)^2) \times 41}{(1 - 0.9897^2) \times 9} = 4.46.$$

This value for the F-ratio is associated with $p < 0.01$ (see the F-ratio tables in Appendix 3), which means that the variances were significantly different at this level of probability and the regression is not likely to be linear. Thus, the multi-target survival curve is not appropriate for these data. The real moral of this story, as with the potatoes, is that it is perfectly possible to have a highly significant regression correlation between two "commodities" even though you

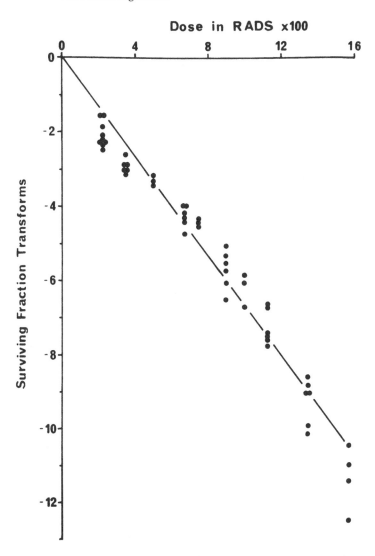

Figure 5.8. Multitarget linear transform applied to cell-survival data obtained from EMT6/M/CC cells grown in tissue culture.

have used a mathematical model for the analysis which bears no relation to the reality. In this case, however, it was possible to prove that the model was inappropriate.

Logistic equation

The final linear transform is of the logistic equation:

$$y = Y_0 \times (1 + \exp(-(\alpha + (\beta \times \log_e(x)))))^{-1}.$$

This can often be used to approximate non-linear data behaviour dependency on the independent variable, as the sets of curves generated by varying the constants Y_0, α and β can assume a large variety of different forms. Dividing through by Y_0, taking the reciprocal of both sides and then subtracting 1.0 from both sides gives

$$\frac{Y_0}{y} - 1 = \exp(-(\alpha + (\beta \times \log_e(x)))).$$

We now take logs to the base e to obtain

$$\log_e\left(\frac{Y_0}{y} - 1\right) = -(\beta \times \log_e(x)) - \alpha.$$

This completes the linear transformation which, again, is in the form of $y = mx + c$. The whole of the expression on the left-hand side is equivalent to y, β is the slope m, and α is the Y-axis intercept c. Hence, by performing regression analysis of $\log_e((Y_0/y) - 1)$ versus $-\log_e x$ we obtain a straight line having slope β which intersects the Y-axis at c, from which get $\alpha = -c$.

Although the non-transformed equation has three constants – namely, Y_0, α and β – the transform is dependent only on Y_0, as the manipulation enables α and β to be obtained as a consequence of the regression. The value of Y_0 is found by successive iteration which maximises the correlation coefficient. We will see how this transformation can be used in Section 10.3.4.

5.2.2   Successive approximation

We considered the following example in Sections 4.4.1 and 4.4.2:

$$y = x^{(k \times 2)} + c.$$

This is a slightly "funny" way of writing this equation, as both components in the exponent, k and 2, are constants. However, when we guessed that the data might be represented by an equation of this particular form, we did not know what the value of the constant in the exponent might be. There are three possible ways we might set about obtaining the best fit between the data and this relationship. First, a log to the base x transform could be performed. We start by rearranging the equation to give

$$y - c = x^{(k \times 2)},$$

and we proceed by logging both sides:

$$\log_x(y - c) = k \times 2.$$

This is not entirely satisfactory, because each y-value would have to be transformed to a different base x. We could do this without any trouble because we know the value of each x, the independent variable, but we do not know the value of c. Thus, we would have to make multiple guesses for c to find the overall best value, which would give us the best average for the constant $k \times 2$. Second, we could differentiate the equation to give

$$\frac{\delta y}{\delta x} = k \times 2.$$

At first sight, this may seem a bit of a neat trick, because when we differentiate a constant, in this case c, we get zero. However, we would now have to estimate the slopes of the curves ($\delta y / \delta x$) in the vicinity of the various x-values, which inevitably will introduce more variation, and we have eliminated c which is something we wish to know!

The final method is to use successive approximation keeping the equation in its non-transformed state, varying k and c and finding the values of these constants which give the minimum sum of squared errors or the minimum $\Sigma \chi^2$. This is shown in Figure 5.9, where Σd^2 and $\Sigma \chi^2$, the "errors", are plotted out as contour maps in panels A and B (respectively) for various values of k and c on the Y- and X-axes, respectively. The two maps show very good agreement for the value of k, both giving the minimum error with $k = 1.0$. However, there is not quite such good agreement for c. Panel A indicates that the sum of squared error is at a minimum with c in the range from 0.05 to 0.1 (hatched area), but the contours are relatively flat in the base of the hole and values of c in the range from -0.2 to $+0.2$ give good fits. Panel B shows $\Sigma \chi^2$ to be at a minimum with c close to 0.1 (hatched area) but a range of from 0.04 to 0.3 gives perfectly satisfactory results. It is apparent from both contour maps that there is relatively more variability in the estimate for c than for k, as the contours are more compressed in the Y-axis direction. This is termed *conditioning,* where here k is better conditioned than c. There is also an interesting feature in the $\Sigma \chi^2$ map: a ridge running slightly inclined away from the vertical in the vicinity of $c = -0.3$ to $c = -0.2$ with a very high local peak. In contrast, no such "fine structure" is seen in the contour map of panel A. This fine structure occurs when the intercept c is sufficiently negative to cause the expected value from the function (the Y-axis value) to be very close to zero for small values of x. Thus, the value of χ^2 calculated as $(O-E)^2/E$ for small values of x can be very large as E tends to zero. This type of structure will never occur when calculating the sum of squared error, as there is no denominator. The sum-of-squared-error map should be used as the means of locating the minimum of a function, but the $\Sigma \chi^2$ map generated simultaneously can be very useful for alerting the investigator to the possibility of "forbidden" or inconsistent domains.

The best-fit values for k and c in this analysis were 1.0 and 0.06, respectively. Table 31 shows results for the model predicted, Y_{expct}, and that observed, Y_{obs}, together with the differences and their squares. We now get the mean deviation as -0.035 with standard deviation of 0.43. This gives Student's $t = 0.181$, which is very much better than for the guesses in Section 4.2.1.

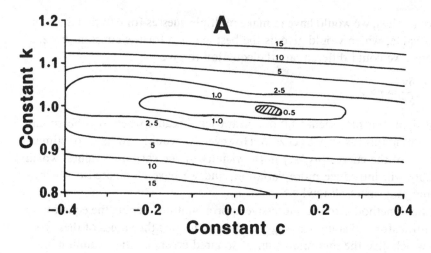

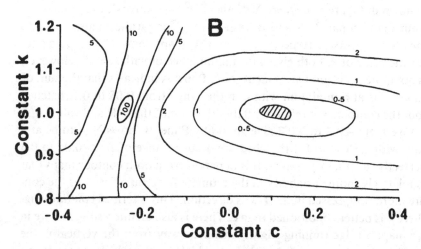

Figure 5.9. Error contours, where $\sum d^2$ and $\sum \chi^2$ are plotted in panels A and B (respectively) for various values for k and c on the Y- and X-axes, respectively.

Table 31

Y_{expct}	Y_{obs}	Difference	d^2
0.31	0.50	0.19	0.036
1.06	0.65	−0.41	0.168
2.31	2.50	0.19	0.036
4.06	4.50	0.44	0.194
6.33	5.50	−0.81	0.656
9.06	9.25	0.19	0.036
Total		−0.21	1.126

5.2.3 Polynomial regression

The polynomial can also be used for non-linear regression analysis. This set of functions has the form

$$y = b_0 + b_1 x + b_2 x^2 + \cdots + b_n x^k,$$

where k is the order of the polynomial and, as with the logistic equation, it can assume a number of different forms depending on the coefficients. If there are n different pairs of points then the curve will pass through all those points (assuming no dispersion) if k, the order of the polynomial, is chosen such that $k = n - 1$.

If there are n pairs of points with the frequency of each (x_i, y_i) pair being f_i, then the total frequency N is given by the summation

$$N = \sum_{i=1}^{i=n} f_i.$$

Let P_i be a point on the curve at x_i, with a y-value Y_i given by

$$Y_i = b_0 + b_1 x + b_2 x^2 + \cdots + b_n x^k$$

$$= \sum_{j=0}^{j=k} b_j x_i^j.$$

If another point Q_i at x_i is distant from the regression curve with an ordinate of y_i, then the deviation D_i of Q_i from P_i is given by $D_i = y_i - Y_i$. Hence, the total sum of squared deviations is

$$NS^2 = \sum_{i=1}^{i=n} f_i (y_i - Y_i)^2.$$

The coefficients b_j $(j = 0, \ldots, k)$ must be chosen so that the sum just shown is at a minimum. This is done by equating the partial derivatives of S^2 associated with those coefficients to zero, so that

$$\sum_{i=1}^{i=n} f_i x_i^j (y_i - Y_i) = 0$$

for $j = 0, \ldots, k$, which gives $k + 1$ so-called normal equations.

If the previous equation is expanded for the second-order system, where

$$Y = b_0 + b_1 x + b_2 x^2,$$

we obtain the following three normal equations:

$$\sum_{i=1}^{i=n} f_i (y_i - (b_0 + b_1 x_i + b_2 x_i^2)) = 0;$$

$$\sum_{i=1}^{i=n} f_i x_i (y_i - (b_0 + b_1 x_i + b_2 x_i^2)) = 0;$$

$$\sum_{i=1}^{i=n} f_i x_i^2 (y_i - (b_0 + b_1 x_i + b_2 x_i^2)) = 0.$$

The coefficients can now be found from the sums of powers of x_i from $\sum x_i^0$ to $\sum x_i^k$ and the sums of products from $\sum y_i x_i^0$ to $\sum y_i x_i^k$, where in this particular case $k = 2$.

In order to simplify things, let all the f_i frequencies be unity; then the above three normal equations on rearrangement become

$$\sum y = b_0 n \quad + b_1 \sum x \quad + b_2 \sum x^2, \tag{5.1}$$

$$\sum xy = b_0 \sum x \quad + b_1 \sum x^2 + b_2 \sum x^3, \tag{5.2}$$

$$\sum x^2 y = b_0 \sum x^2 + b_1 \sum x^3 + b_2 \sum x^4. \tag{5.3}$$

A further simplification is also possible if there are equal increments δ between each x-value, in which case a change of variable is used to eliminate the odd powers of x. This uses a transformation u such that $u = (x - m)/\delta$, where m is the mid-point of the x-values. This transform will be demonstrated for both odd and even numbers of points in the x-array.

Let there be six x-values at unit intervals, where for $i = 1, \ldots, 6$ $x_i = 1, 2, 3, 4, 5$ and 6. The mid-point m in this array is 3.5, hence the interval δ between the mid-point and the two nearest values is 0.5. We now carry out the change of variable (transformation) where $u = (x - m)/\delta$. This gives $u = (x - 3.5)/0.5 = 2x - 7$, which equates the mid-point of the x-values to zero. The new units of u are shown in Table 32, together with the original x-values and associated y-values, where $y = x^2$ plus the sums and product terms needed to calculate the coefficients.

It can now be seen that, with this new unit system, all sums of odd powers of u must be zero, which is the reason for the change of variable. The various summed values can now be substituted into equations (5.1)–(5.3) to give

Table 32

x	u	y	u^2	u^4	uy	$u^2 y$
1	−5	1	25	625	−5	25
2	−3	4	9	81	−12	36
3	−1	9	1	1	−9	9
4	1	16	1	1	16	16
5	3	25	9	81	75	225
6	5	36	25	625	180	900
Total		91	70	1414	245	1211

$$91 = b_0 6 + \quad 0 \quad + b_2 70, \tag{5.4}$$

$$245 = \quad 0 \quad + b_1 70 + \quad 0, \tag{5.5}$$

$$1211 = b_0 70 + \quad 0 \quad + b_2 1414. \tag{5.6}$$

The value of b_1 is now obtained directly from equation (5.5) as $b_1 = 245/70 = 3.5$. Equations (5.4) and (5.6) are solved simultaneously to give $b_1 = 0.25$ and $b_0 = 12.25$, yielding the regression equation

$$y = 12.25 + 3.5u + 0.25u^2.$$

We can now change back into the original units by substituting $2x - 7$ for u to obtain

$$\begin{aligned} y &= 12.25 + 3.5(2x - 7) + 0.25(2x - 7)^2 \\ &= 12.25 + 7x - 24.5 + 0.25(4x^2 - 28x + 49) \\ &= 12.25 + 7x - 24.5 + x^2 - 7x + 12.25 \\ &= x^2. \end{aligned}$$

We now consider just the first five points, so that the mid-point in the array of x-values is 3 and the interval between this mid-point m and the next value is unity; that is, $\delta = 1.0$. The u-transformation is again $u = (x - m)/\delta$, hence $u = x - 3$ which gives the various values in Table 33 corresponding to those in Table 32.

Again, the odd powers of u are zero, and the various summed values in Table 33 are substituted into equations (5.1)–(5.3) to give

$$55 = b_0 5 + \quad 0 \quad + b_2 10,$$

$$60 = \quad 0 \quad + b_1 10 + \quad 0,$$

$$124 = b_0 10 + \quad 0 \quad + b_2 34,$$

Table 33

x	u	y	u^2	u^4	uy	$u^2 y$
1	-2	1	4	16	-2	4
2	-1	4	1	1	-4	4
3	0	9	0	0	0	0
4	1	16	1	1	16	16
5	2	25	4	16	50	100
Total		55	10	34	60	124

from which we obtain $b_0 = 9$, $b_1 = 6$ and $b_2 = 1$. Our regression curve now becomes

$$y = 9 + 6u + u^2,$$

which on substituting for u gives

$$y = 9 + 6(x-3) + (x-3)(x-3)$$
$$= 9 + 6x - 18 + x^2 - 6x + 9$$
$$= x^2.$$

Hence, with the appropriate u-change of variable for both odd and even numbers of points, we obtain the regression curve $y = x^2$.

Now let us return to the data of Figure 4.5, where the best fit to the equation $y = x^{(k \times 2)} + c$ was found in Section 5.2.2 by successive approximation with $k = 1.0$ and $c = 0.06$. The x-values occur at 0.25 intervals with the mid-point at 1.75, hence $\delta = 0.5$. This gives $u = (x - 1.75)/0.25 = 4x - 7$, and the associated sums and products are shown in Table 34. Substituting these various terms in the normal equations and simplifying gives

$$y = 0.2238 - 0.2056x + 1.0432x^2. \tag{5.7}$$

The regression y-values (Y_{expct}) and the observed (Y_{obs}), with their differences and squares, are shown in Table 35. The mean difference was 0.002833 with standard deviation, after Bessel's correction, of 0.469386. This gives Student's $t = 0.014784$, which is considerably better than obtained by successive approximation in Section 5.2.2. The best-fit curve in the latter was $y = 0.06 + x^2$, which should be compared with equation (5.7). The two are similar, but with the difference that equation (5.7) has an x-to-the-power-of-1- (i.e. x-) term included. The equation used for the successive approximation did not include this possibility. When I generated the "data" for this particular example I based it on the equation $y = x^2$ and added in some random variation. However, introduction of the latter resulted in the equation on which the data were originally based

Table 34

x	u	y	u^2	u^4	uy	u^2y
0.5	-5	0.50	25	625	-2.50	12.50
1.0	-3	0.65	9	81	-1.95	5.85
1.5	-1	2.50	1	1	-2.50	2.50
2.0	1	4.50	1	1	4.50	4.50
2.5	3	5.50	9	81	16.50	49.50
3.0	5	9.25	25	625	46.25	231.25
Total		22.90	70	1414	60.30	306.10

Table 35

Y_{expct}	Y_{obs}	Difference	d^2
0.382	0.50	−0.118	0.0139
1.061	0.65	0.411	0.1689
2.263	2.50	−0.237	0.0562
3.985	4.50	−0.515	0.2652
6.230	5.50	0.730	0.5329
8.996	9.25	−0.254	0.0645
Total		0.017	1.1017

being no longer the most appropriate for the analysis, even though the $+c$ constant was added. Thus, if you suspect that an equation of the form $y = x^2$ might fit the data, don't use it; use instead the polynomial regression analysis described here. If the equation is $y = x^2$ then the polynomial analysis will tell you that $b_0 = b_1 = 0$ and that $b_2 = 1$, which reduces the polynomial to $y = x^2$.

6

Flow cytometric sources of variation

Before we see how the concepts outlined in previous chapters can be used we should consider the reasons why they might have to be used. Some of the sources of variation in making a measurement with a ruler were discussed in Section 2.2. It is very important to appreciate where sources of variation in flow cytometry might be encountered, because variation gives rise to distributed data from the population being analysed. Flow cytometers are somewhat complex, and sources of variation can enter at every step in making a quantitative recording. We have already seen that the measurements we make are indirect (Section 2.2) in that there is conversion of light energy into electrons. But the process starts even further back down the line. Initially, the sample is prepared and stained; next it flows through the laser focus from which the fluorescence is elicited; this is then collected and focused onto the photomultiplier; converted to a voltage as already described; amplified and subjected to signal processing, including analogue-to-digital conversion; and finally stored electronically. These are just the physical processes, all of which can introduce variation, but there are also biological factors to be included. Some of these factors are more important than others, but each will be considered.

6.1 Preparation and staining

There are usually many steps in the preparation and staining processes. In solid tissues a disaggregation step is required to produce a single cell suspension, and this frequently involves enzymatic treatment. This, as well as some of the other preparative and staining steps, induces variation.

6.1.1 Quenching

I'm starting with this topic as I need to refer to it in the next section. Quenching is defined as a reduction of fluorescence yield due to loss of the absorbed excitation energy by any pathway other than light emission, and there are a number of mechanisms. First, the excited fluorophore may revert to the ground state by non-radiative energy transfer within the fluorophore–carrier complex. Second, the energy may be dissipated to the microenvironment, a process that is frequently pH dependent. Third, bleaching may occur if the energy from a second photon is absorbed by a fluorophore in the excited state before it has discharged the energy absorbed from a first photon. This can result in a number of changes, ranging from *photolysis* – complete disruption of the molecule with the formation of two or more new species – to more minor alterations in electronic configuration. The latter are usually reversible as far as fluorescence is concerned, and everyone who has used a fluorescence microscope will have observed the bleaching of fluorescein fluorescence and its recovery after the illumination is turned off for a few minutes.

Finally, resonant energy transfer may occur. In this particular type of quenching the excitation energy absorbed by one molecule is resonantly transferred to a second molecule. This is a non-radiative process: the first molecule (the donor) returns to its ground state, leaving the second molecule (the acceptor) in an excited state which can then revert to the ground state by any of the mechanisms available, including light emission. Two conditions must coexist for this process to occur. First, the emission energy spectrum of the donor must overlap the absorption spectrum of the acceptor; second, the molecules must be very close to each other. The efficiency of resonance energy transfer is inversely proportional to the sixth power of the distance between the molecules, and very little energy can be transferred beyond 60–70 Angstroms (Stryer 1978; Jovin 1979).

6.1.2 Stoichiometry

The definition of this term appears in the Chambers Twentieth Century Dictionary (1976) as "the branch of chemistry that deals with the numerical proportions in which substances react" (from Greek *stoicheion,* an element; *metron,* measure). In flow cytometry, stoichiometry refers to the state where the quantity of fluorescent light emitted from a dye–ligand complex is directly proportional to the quantity of the target molecule. Thus all, or very nearly all, the free binding sites should be complexed with dye either directly or indirectly.

There is a potential conflict in definition of terms here. At this point we are considering stoichiometry, and the term "directly complexed" refers to the state where the dye is bound directly to the ligand as in DNA staining. The term "indirectly complexed" refers to staining as in immunofluorescence work, but it will be appreciated that techniques in the latter also include both direct and indirect methods. In order to avoid ambiguity I will use the term "dye–ligand complex" and apply it to all such interactions.

The emitted fluorescence from the dye–ligand complex is directly proportional to both the number of dye molecules bound and the quantity of exciting light actually absorbed by those molecules in the complex. The processes involved are shown in Figure 6.1, represented as a biochemical process such as an enzyme reaction. It may seem a little strange at first to represent the emission of fluorescence as one would normally represent an enzyme reaction. However,

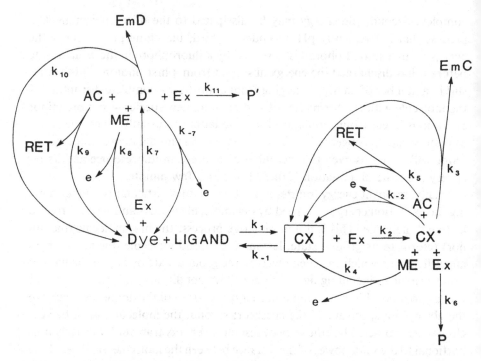

Figure 6.1. Representation of the processes involved in dye–ligand binding, absorption of light, fluorescence emission and quenching (see text for explanation).

the two processes have similarities, particularly if you allow your mind to consider them simply as the manifestations of energy transfer systems that they are.

The various components represented in Figure 6.1 and the processes affecting these components are as follows. DYE and LIGAND represent unbound molecules, which are the available and free binding sites. These may combine with rate constant k_1 to form the dye–ligand complex denoted by CX. Once CX is formed it could dissociate and revert to free dye and unbound ligand in the "back-reaction" with rate constant k_{-1}. If the system is an antigen–antibody reaction then the forward-rate constant k_1 is usually of considerably greater magnitude than the back-rate constant k_{-1}. However, k_{-1} will never be zero, so there is always a definable probability that the dye–ligand complex can dissociate. When light excites this system a number of processes can take place. CX can "combine" with exciting light (Ex) with rate constant k_2 to form the excited complex CX*. This can then emit fluorescence (EmC), which is what we want to measure, with rate constant k_3. However, four possible quenching processes involving CX* can also take place. First, the excited complex may revert to the non-excited state CX by non-radiative energy transfer within the bound molecular complex with rate constant k_{-2}. Second, CX* may also dissipate energy to the microenvironment (ME) by non-radiative transfer with rate constant k_4. Third, resonant energy transfer (RET) may take place with rate constant k_5 if there is an appropriate acceptor molecule. Finally, CX* may disintegrate due to photolysis (P), absorbing further photons whilst in the excited state, with rate constant k_6.

In immunofluorescence staining there should be very little "free" dye available after the antibody washing steps. However, this is not generally the case with DNA staining, and the free dye may also undergo the same quenching processes just described.

6.1.3 Binding-site modulation

A number of types of binding-site modulation can occur, but operationally these can be grouped into two major categories: those involving immunofluorescence staining and those involving direct binding of dye to ligand (e.g. dye–DNA interactions).

Any form of fixation used to cross-link and/or agglutinate proteins in the preservation process can damage epitopes and inhibit subsequent binding of an antibody. This is highly dependent on the epitope being probed, and cannot be predicted in advance. The only way to find out is to suck it and see. Essentially, there are two types of fixatives, those which cause agglutination (alcohols, acetone) and those which cross-link (aldehydes); see the review by Hopwood (1985). If one type of fixative reduces antibody binding (van Ewijk et al. 1984) it is always worth trying a different type. However, if one category fails it is quite possible that another will also fail; van Ewijk et al. (1984) have reported that this problem can be solved with some antibodies by staining before fixing. The latter may need to be carried out if the assay being performed requires that a second polar fluorochrome be used to probe an intracellular molecule – for example, propidium iodide for DNA.

Epitopes of cell surface determinants may also be masked or partially masked if these are close to the cell membrane. This type of problem has been reported for antibody staining of the efflux pump gp[170] by Cumber et al. (1990), who also report that increased binding can be obtained by pre-treating cells with neuramidase, which de-sialates the external membrane. The inference here was that heavily sialated membranes were causing steric binding hindrance.

Similar types of problems may attend DNA staining. Chromatin-associated proteins may reduce the accessibility of dyes to binding sites either by physically covering the sites or by modifying nucleic acid conformation (Angerer and Moudrianakis 1972; Brodie, Giron and Latt 1975; Fredericq 1971). It is of course always possible to control dye concentration but it is not always possible to control the number of accessible binding sites in DNA, and staining techniques using acridine orange have been specifically designed to remove basic histones to increase the number of accessible binding sites (Darzynkiewicz et al. 1975, 1977a; Traganos et al. 1977). The state of chromatin condensation and associated DNA denaturability can make profound differences in the number of accessible binding sites, and this has been exploited by Darzynkiewicz et al. (1977b,c) for many years to distinguish between G2 and mitotic cells. In tissues that have been fixed there may be permanent cross-linking in DNA and its associated proteins due to the fixation process which cannot be "unfixed", and this too can mask potential binding sites. Prolonged formalin exposure, particularly in non-buffered saline, and mercury-containing fixatives can give rise to problems of this nature. The time interval between interruption of the blood supply to the surgical specimen and either fixation or processing can also modify the available binding sites.

Cells contain endonucleases and proteolytic enzymes which digest DNA and nucleoproteins respectively. Normally, any intracellular proteolytic enzymes are maintained in an inactive state or are contained in subcellular compartments, but with cell death these enzymes are released and generally "chew up" the cells from the inside. This inevitably degrades DNA, reduces the number of binding sites and degrades the quality of the data obtained. Hedley (1989) has stressed the importance of adequate fixation of paraffin-embedded material. Clearly, if fixation is not adequate then there will be tissue degradation both before and during the embedding process. In the latter the temperatures reach above the DNA melting point, and if tissue is inadequately fixed then there will be temperature-dependent DNA degradation. Feichter and Goerttler (1986) have identified inappropriate storage conditions of paraffin-embedded biopsies as a cause of artefact. Prolonged (years, not specified) central heating temperatures were sufficient to degrade tissue contained in wax blocks. These authors have also identified tissue damage before fixation as a source of artefact. Biopsies that had been stored frozen at temperatures between $-40°C$ and $-60°C$ which were thawed, fixed and subsequently embedded gave uninterpretable DNA histograms.

6.2   Fluorescence excitation

As previously stated, the quantity of light emitted from a dye–ligand complex is proportional to both the quantity of that complex and the excitation intensity. Potential sources of variation in the former were addressed in the previous section and the latter are discussed in this section.

6.2.1   Light-source stability

Theoretically, light-source instability could contribute as a source of variation. However, the vast majority of flow systems use lasers and these are now very stable, but they should be operated in the light-stabilised and not the current-stabilised mode. Under these conditions most lasers will give a constant output to within ±0.5% over many hours of continuous operation. Instruments using arc-lamp illumination usually have an electronic feedback loop that corrects for any illumination instability. This works well as long as the intensity does not vary by more than about ±5%. However, towards the end of the lifetime of a mercury arc (about 150–200 hours) the intensity can vary by considerably more than this, and the lamp must then be replaced.

6.2.2 Illumination constancy

The beam from a laser can assume various so-called Transverse Emission Modes (or TEM for short) which depend on a number of factors including plasma-tube construction and mirror alignment. These modes are subscripted as 00, 01, 01*, 10, 11 and so on depending on the segmentation of the beam. The intensity of the beam is Gaussian distributed in TEM_{00}, and it is conventional to describe the width of a beam as the diameter across the beam where the irradiance falls to 0.1353 ($1/e^2$) of that at the peak. Let us now assume that a typical 1.2-mm diameter beam is focused down to a 50-μm diameter spot at the $1/e^2$ irradiance width, and that a core stream diameter of 15 μm containing 10-μm cells is passing through the center of the focused beam in the most intense section. This is depicted in Figure 6.2, with cells A and B in the center and at the periphery of the core, respectively. Cell A, in the ideal position, has a 6% variation in light intensity across it. Cell B, in the worst position at the core periphery, suffers a 15% intensity variation. We can see that very minor instability in the core position, one of as little as ±2 μm, could make profound differences to the illumination intensity experienced by individual cells. With this degree of core instability the variation could easily amount to 40% as depicted in Figure 6.2, which is a reasonable to-scale representation of the reality.

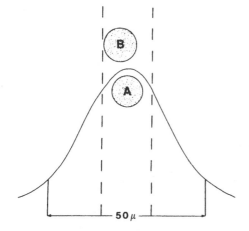

Figure 6.2. A 15-μm diameter core passing through the center of the focused Gaussian beam in the most intense section and containing two 10-μm cells. Cell A, in the ideal position, has a 6% variation in light intensity across it. Cell B, in the worst position at the core periphery, suffers a 15% intensity variation. Core position instability of as little as ±2 μm could give rise to a 40% illumination variation.

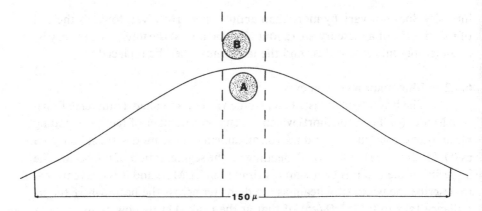

Figure 6.3. Crossed cylindrical–lens pair illumination with compression of the focal spot from top to bottom to form an oval sheet 150 μm wide and 4–7 μm deep. A 15-μm diameter core containing two cells, directly analogous to that in Figure 6.2, is shown. The potential variation in light intensity across a 10-μm cell is now less than about 2%, allowing considerable latitude for any core positional instability.

Two options are available to overcome this potential variation in illumination intensity. First, a larger spot size could be used to "spread out" the highest-intensity region with respect to the core, but this also reduces the light flux which is not desirable. The second is to use a crossed cylindrical lens pair, focusing where the focal spot is compressed from top to bottom and spread laterally to form an oval sheet. The intensity profile is still Gaussian distributed in whichever radial plane the beam is "sliced", but the standard deviation is "tightest" in the vertical plane. Figure 6.3 shows a representation of the focal-spot intensity obtained with a crossed cylindrical lens pair that spreads the beam to 150 μm horizontally together with a 15-μm diameter core, directly analogous to that in Figure 6.2. The potential variation in light intensity across a plane of a 10-μm cell is now less than about 2%, so there is now considerable latitude for any core positional instability.

6.2.3   Hydrodynamic instability

This is potentially one of the most important causes of excitation variation, and we have seen in the previous two figures the relationship between the position of cells in the core and the intensity profile of the excitation beam. Obviously, core instability will induce variation in excitation intensity. The most frequent cause of hydrodynamic instability is badly prepared samples containing junk. In my experience the latter has included shredded cells (common), fibers from plugged Pasteur pipettes, animal hair and slivers of plastic scraped off from petri dishes when cells are harvested clumsily with glass pipettes. In order to minimise this problem all samples should be filtered through a 35-μm nylon mesh before introduction into the instrument.

6.2.4 Bleaching

This phenomenon can occur at relatively low laser light powers with some fluorophores. It will contribute as a potential source of variation particularly if there is core instability (messy samples) that induces illumination inconstancy. Cells in the higher light-flux region will be bleached to a relatively greater extent than cells in a lower light-flux region.

6.3 Optical design

Optical design influences the total quantity of light reaching the photodetector. The latter have not only a finite operating range but also a finite range over which the response is linear. Moreover, there is always a "dark current" which will restrict the range. Hence the possibility of increasing this background dark current must be minimised, and good optical design can help in this aspect.

6.3.1 Collection efficiency

Because cells are small and are interrogated for a few microseconds at most, it is clearly desirable to collect as much light as possible. Most of the older flow systems have a light-collection efficiency of about 5% but in more modern varieties the efficiency has been increased to between 15% and 25%, and this would seem to be universally desirable. However, as with all things, there are two sides to the coin. Collecting more light necessarily means that more scattered and background light as well as fluorescence will be collected. Thus, in order to obtain an increased signal-to-noise ratio, the extra background light must be eliminated completely (which is usually not possible) or reduced to a minimum. Moreover, we must ensure that the extra wanted fluorescent light collected does not start saturating the photodetector and thus produce a non-linear response. Occasionally this will require that neutral density filters be placed in front of the detector. If the light-collection efficiency of a system is increased by a factor of, say, three then we have to reduce the probability of background light entering the fluorescence detector by this same factor to obtain a threefold increase in the signal-to-noise ratio.

6.3.2 Fluorescence of filters

Some materials used in the manufacture of optical filters are fluorescent. This applies particularly to some of the older coloured glass filters. If such filters are being used and there is a relatively large quantity of exciting light being scattered towards the light-collection optics, it is possible to get fluorescence from the filter which the instrument will associate with the cell. The quantity of light scattered towards the detector is proportional to the cell's physical characteristics, which vary from cell to cell. Hence this is a potential source of variation.

6.3.3   Filter combinations

The performance of any filter system depends on some form of probability function, and there is always some chance that an exciting photon will enter the fluorescence detector and be scored erroneously as fluorescence. This phenomenon is frequently used in setting up an instrument with non-fluorescent control cells by increasing the photomultiplier (PMT) high-tension voltage until the scattered exciting light from each cell "breaks through" the filter system and can be seen by the fluorescence detector. The sample containing labelled cells is then run and the two data sets are compared.

Modern photomultipliers and electronics are capable of amplifying a signal by many orders of magnitude, and this has important consequences for the measurement of very low levels of fluorescence and for determination of sensitivity and detection limits. This applies particularly when the excitation light is close to the lower wavelength limit above which fluorescence is measured, and where the latter is less intense than the excitation light scattered towards the fluorescence detector. Hence, a major component of that which is scored as fluorescence could in fact be scattered light. Perhaps the best example is fluorescein excited with the blue 488-nm line from an argon laser, where green fluorescence is measured above about 515 nm. The filter system should contain a dichroic mirror centered at about 510 nm to reflect out scattered light from the path to the green fluorescence detector, and the latter should also be guarded by a 515–560-nm band pass filter. The problem tends to be compounded by the compulsion to increase laser power for detection of low-intensity fluorescence, as the yield of the latter tends to saturate due to bleaching while the quantity of scattered light continues to increase linearly.

Careful design of filter combinations is required to minimise this as another potential source of variation. In very high-efficiency light-collection instruments an extra dichroic mirror can be introduced into the filter system to further decrease the probability of a 488-nm photon entering the green detector.

6.4   Electronic factors

The electronics of flow cytometers are becoming increasingly sophisticated and are very unlikely to give rise to any significant degree of variation. However, an electronics section is included here because features within the electronics can be used very effectively in quality control to exclude debris, clumps and coincident events from the data base.

6.4.1   Triggering and thresholds

The electronics are designed so that signal processing begins only at a set voltage, or threshold, above the background baseline level. This is important, as there has to be some criterion which sets the electronic wheels in motion and tells them when to start turning. You cannot have the electronics in action

Table 36. *Poisson analysis of coincidence*

Peak	Observed channel	Expected channel	Observed frequency	Poisson frequency
Singlet	19	(19)	97.1%	98.2%
Doublet	39	38	3.9%	2.7%
Triplet	59	57	0.023%	0.048%

all the time because you would never be able to discriminate between rubbish, random fluctuations and reality (cells, hopefully). It is also important to be able to specify a master triggering channel – particularly for multiparameter work (which in this context means more than 1) – that is capable of controlling the electronic initiation.

6.4.2 Coincidence correction

Coincidence is usually to be avoided like the plague, but it can be used to check the linearity of the instrument. Microbeads are run through the instrument at a high concentration and the Poisson probability of obtaining coincidence of two and three events in the sensing volume can be calculated. The peak positions of and proportions in the singlet, doublet and triplet distributions are then calculated. The data in Table 36 show the results of one such experiment, carried out a long time ago using a Bio-Physics 4800A cytofluorograf (Watson 1977a). These results demonstrate that the response of the instrument was linear, even though there were small discrepancies in the observed and expected peak channel positions and in the observed and Poisson predicted proportions in the distributions. In fact, these results were found to be even better when I later discovered that there was a single channel offset on the ADC (see Section 6.4.5), so that the observed peak channels should have been 20, 40 and 60 and not 19, 39 and 59.

There are two ways in which the electronics can be used to help in coincidence correction, and each uses time slightly differently. First, the computer used for data acquisition can be programmed to calculate the average time interval between arrival of cells at the analysis point based on the rate at which the digitised signals are being presented to it for storage. It can then calculate the Poisson probability of coincidence and set the data-acquisition logic so that only events spaced at greater than specific intervals are recorded. Thus, if events are arriving "too close" to each other they are both ignored.

The second method uses a double–electronic threshold triggering system. Figure 6.4 shows pulses from two closely spaced cells passing through the beam. Threshold 1 (T1) is set lower than threshold 2 (T2), and nothing at all will happen unless the voltage from the photodetector exceeds T1. This voltage

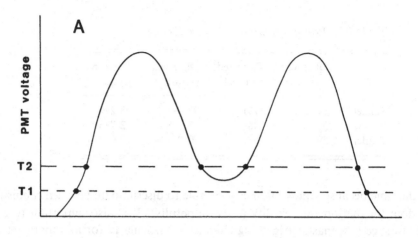

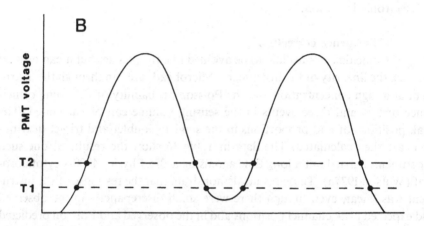

Figure 6.4. Double–electronic threshold triggering system used to identify closely spaced pulses.

"enables" the electronic acquisition system, which begins to integrate the area under the pulse, starts timing the width of the pulse and records the peak height when this is reached. In the top panel the voltage drops below T2 as the first cell passes out of the beam, but does *not* drop below T1 before rising again to cross T2 as the second cell enters the beam. This aborts the data-acquisition logic, which is then re-set and the recording is not completed. However, in the lower panel the voltage drops below T1 before it rises again through T2. In this case a recording is made: the width of the pulse is timed when the voltage drops below T1, and pulse height, width and area are digitised and stored.

This system is similar to traffic lights controlling vehicle flow at a road junction. If only the amber light is showing you know that red is coming up next and you must stop. If red and amber are showing then green is coming up next and you can prepare to accelerate. With the double-threshold system the sequence

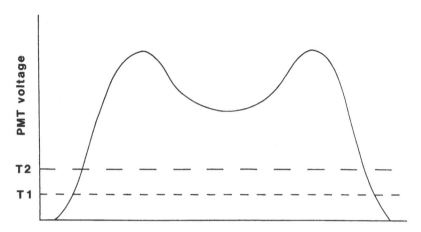

Figure 6.5. Illustration of two closely spaced pulses that would not be recognised by the double-threshold system.

T1, T2, T2, T2 means stop – events are too closely spaced. In contrast, the sequence T1, T2, T2, T1 means go and the data are acquired.

There is a third possibility, shown in Figure 6.5, where the threshold sequence would be T1, T2, T2, T1 and the data would be recorded even though this clearly is a double event. However, this sequence can be excluded after acquisition by using pulse-shape analysis, which is particularly necessary for some types of preparations such as nuclei extracted from paraffin wax–embedded archival material. These inevitably contain a variable quantity of debris, clumps and nuclear fragments as well as intact single nuclei. After DNA staining with propidium iodide, any small debris and large clumps can be gated out on a combination of forward and 90° scatter signals. Large fragments of nuclei and small clumps not identifiable in the forward versus 90° light-scatter data space can be identified and gated out using the shape of the pulse from the red (DNA) photomultiplier, which is usually the master triggering detector in these types of assays. This involves analysis of the pulse height, width and area under each pulse (PAW analysis; see Watson, Sikora and Evan 1985), where the ratio of (width × height)/area should be constant to within narrow limits.

The method is illustrated in Figure 6.6, which shows individual pulses from a single cell (left panel) and a clump of two cells (right panel), similar to Figure 6.5. The boxes surrounding each panel represent the area obtained by multiplying the height of each pulse by the width, time of flight through the beam. Naturally, this analysis can be used only in partial slit–scan systems using crossed cylindrical–lens pair focusing, where peak pulse height, width and area are digitised. We can see by inspecting the two panels that the integrated area under each pulse, which is shaded, occupies a different proportion of the boxes defined by multiplying height by width. Because the beam illumination intensity is Gaussian distributed, nuclei that are approximately spherical will give single-cell

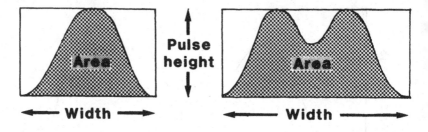

Figure 6.6. Pulse-shape analysis of height, width and area under each pulse (PAW analysis), where the ratio of (width × height)/area should be constant to within narrow limits. This can be used to discriminate between fragments and clumps not identified by the double-threshold system.

pulse shapes which approximate to a Gaussian profile no matter what the size. Thus, single-cell events will give a constant for (height × width)/area. Doublet clumps (right panel of Figure 6.6) and fragments of nuclei will not give a constant value for this derivative, but rather a spread in values which are unlikely to be the same as the values from intact nuclei. Thus, if the majority of the population (i.e. more than 50%) is composed of single nuclei, these can be identified and the abnormal pulse shapes excluded.

6.4.3   Linear amplifiers

The output signal from a linear amplifier is directly proportional to the input signal, where for each particular amplifier the output is larger by a defined quantity, the gain, with respect to the input. In some instruments the gain of linear amplifiers can be varied, which increases their measurement range. Let us suppose the amplifier has a maximum output of 10 volts which is directed through to a voltmeter with a scale of 1 to 1000. The 10-volt maximum signal will have a scale reading of 1000, and signals of 5.0 and 2.5 volts would appear on the scale at 500 and 250, respectively; that is, the response is linear. This results in the coefficient of variation (CV), defined as the standard deviation divided by the mean ($\sigma/\bar{x}$), remaining constant with signal magnitude.

6.4.4   Log amplifiers

Log amplifiers were introduced to increase the dynamic range within the instrument; they operate on logarithmic scales of either 3 or 4 decades. If we take a 4-decade log amplifier with a 10-volt maximum output and again feed the output into our voltmeter, we will still get a scale reading of 1000 for 10 volts of output. However, as this is a 4-decade amplifier, each tenfold decrease in the input signal will give an amplified signal that decrements the maximum output range by 1/4. Thus, input signals amplified to 1.0, 0.1 and 0.01 volts will give voltmeter scale readings of 750, 500 and 250, respectively. A comparative representation of the outputs from a linear and a log amplifier are shown in

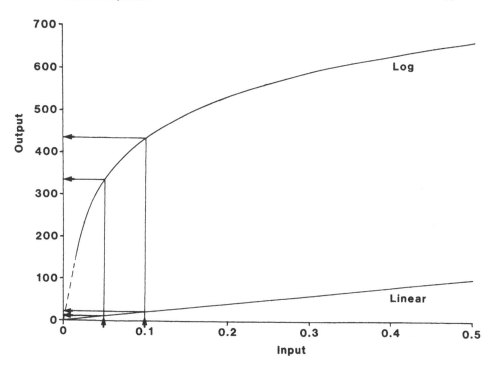

Figure 6.7. Comparison of the outputs from linear and log amplifiers.

Figure 6.7, where the input is in millivolts and the output is in volts × 100. Figure 6.8 compares histograms obtained with each type of amplifier.

Log amplifiers are designed to "expand" the smaller signals, which in turn "compresses" the larger signals. This has an important consequence for data analysis, because these amplifiers produce not a constant CV but rather a constant standard deviation and hence a constant variance.

6.4.5 Analogue-to-digital conversion

The amplified signal, whether from a log or a linear amplifier, must now be converted to a whole number which can then be stored electronically. The analogue-to-digital converter (ADC) has a fixed range, and in flow cytometry this is usually from 0 to 255 inclusive or from 0 to 1023. These are respectively termed 8-bit and 10-bit resolution ADCs.

A description of the workings of an ADC is not in order here, but it is important to know that they can be set up relatively arbitrarily. Thus, the origin can be offset either to the left or to the right. Furthermore, they can also be set to ignore any signals greater than their maximum, be this 255 or 1023. The latter is not good practice, however, and any signals of 255 and above in an 8-bit ADC should be scored as 255. Likewise, with a 10-bit ADC any signals equal to or greater than 1023 should be scored as 1023 and not ignored.

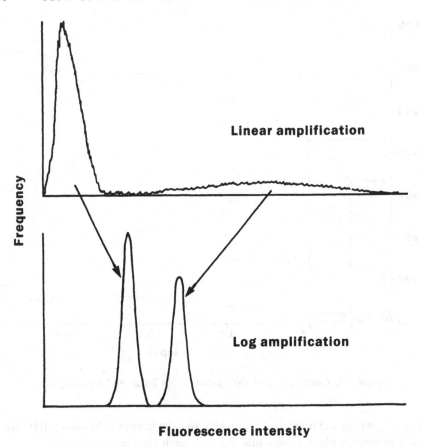

Figure 6.8. Comparisons of histograms obtained with linear and log amplifiers.

It is not generally appreciated that the analogue-to-digital conversion step following linear amplification will introduce a positive skew into the distribution of the parameter being measured. Thus, if a given signal in a 1024 ADC is scored with a value of 100 then we need a minimum increase of 1% in another signal for it to be scored as 101. However, if a signal has a value of 1000 we need an increase of only 0.1% for a signal to be digitised as 1001. Thus, the resolution increases with the magnitude of the signal.

The effect on standard deviation (SD) of a distribution is as follows. Consider a distribution with a SD of 5 and mean of 100. This gives a CV equal to 0.05. Now, for a second distribution with SD of 5 and mean of 1000 the CV is 0.005. If, however, the first distribution (mean = 100, SD = 5) had been recorded with the photodetector gain increased by a factor of ten, then the mean would appear in channel 1000. A cell previously recorded in channel 101 would now appear in channel 1010. Hence, the SD would be 50 with the CV constant and maintained at 0.05. However, there is a limitation to the constancy of the CV.

Consider a distribution with mean of 100 and CV of 0.5%. About 70% of the distribution would be scored in channel 100. However, if this same distribution had been scored in channel 10, the whole of the distribution would have been scored in that channel and the CV would be about 2%. Thus, the CV is constant only above the point at which the ADC step can resolve the dispersion in the distribution.

If there is a large spread (CVs > 0.2) in a measured parameter from a population, the distribution will cover a large proportion of the digitisation steps in the ADC. Thus, cells that elicit a small response will have a smaller local SD than those cells with a large response, which introduces an artefactual positive skew into the recorded data and effectively represents a source of variation. This also applies to data sets with relatively low CVs, but in these cases the effect is less apparent.

6.4.6   Compensation

Electronic compensation is designed to correct for breakthrough of fluorescence from one fluorophore into the analysis window of another and to reduce noise (Steinkamp 1983). Examples include fluorescein (green) into the orange/red photomultiplier quantitating light from rhodamine or phycoerythrin fluorescence, and propidium iodide DNA fluorescence (red) into the green detector quantitating fluorescein. Figure 6.9 shows an example of the last of these, where propidium iodide–stained DNA (red fluorescence) was being scored on the abscissa versus a fluorescenated monoclonal antibody-probed nuclear-associated antigen (green) on the ordinate. The 1-dimensional histograms associated with the two measurements are shown adjacent to the respective axes. These data were from control samples, where panels A and C represent the phosphate-buffered saline (PBS) controls (no fluorescein) and where panels B and D were obtained with added second antibody (fluorescenated) but no first antibody. The instrument was set up to record the G1 DNA peak in panel A at channel 200 on the abscissa. The gain on the green channel was then increased so that the short wavelength tail from the PI–DNA emission spectrum broke through the green band pass filter to record the G1 peak at about channel 25. Clearly, there is a direct relationship between the two measurements. In panel B there is a slight increase in the green signal due to non-specific binding of the fluorescenated second antibody. Compensation was carried out on the data of panel A and the result of this is shown in panel C. The same compensation was also carried out on the data in panel B and these results are shown in panel D, which is the background signal above which any subsequent specific green signal due to first antibody binding would have to be measured.

Figure 6.10 shows a schematic of the processes involved in a differential amplifier to correct for this artefact. In this example we are compensating on the green channel for red breakthrough. The red signal (pulse B) is input to a proportional pulse divider (PPD) which gives two output pulses, C and D. These

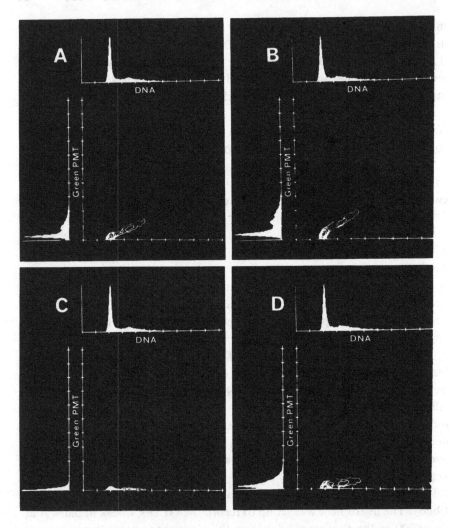

Figure 6.9. Breakthrough of the short wavelength tail from propidium iodide-stained DNA (red fluorescence) into the green channel, panel A. This has been compensated for in panel C. Panels B and D show comparable data for a fluorescence control sample containing fluorescenated second antibody but no first antibody.

always sum to the value of the input pulse, but the ratio of the outputs can be varied; pulse C, which will be used for the compensation, is adjusted to be the same size as A, the green background signal. Pulse C is now amplified by a factor of two and divided equally into two (AD). This gives two output pulses (E and F) each of which is the same size as the input pulse C. One of these pulses, F, is now added back to the divided red pulse D (PA), which restores this to its original size, B. Compensation on the green channel is effected by

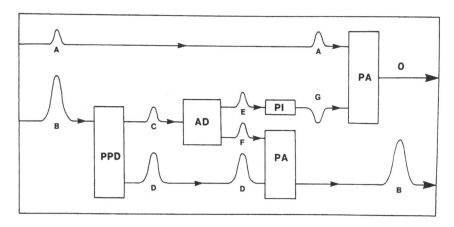

Figure 6.10. Block diagram illustrating the processes involved in differential amplification to correct for breakthrough from one channel into another.

inputting pulse E into a pulse inverter (PI in Figure 6.10), which turns it upside down to give the inverted output pulse G. This is now added to the green input pulse A (PA) to give an output of zero. In the examples shown in Figure 6.9, the background signal in the green channel due to red (PI–DNA) breakthrough has been eliminated, although there has been no change in the DNA histograms (red signals).

6.5   Biological variables

The sources of variations considered to date are dependent on either sample preparation or features of the instrumentation. We now consider sources of variation which are primarily dependent on biological factors.

6.5.1   Autofluorescence

A number of cellular constituents, including NADH and tyrosine, fluoresce with the exciting wavelengths used in flow cytometry. Moreover, fixation with certain aldehydes, gluteraldehyde in particular, can induce considerable cellular fluorescence. This is not infrequently in the high blue/green region of the spectrum and can raise the background above which fluorescein has to be measured. In assays where the fluorescein fluorescence is very "bright" this autofluorescence is of no consequence, but in assays where the specific fluorescence is "weak" a considerable fraction of the variation in the population may be due to autofluorescence.

6.5.2   90° scatter

Some cells (e.g., polymorphonuclear leukocytes in general and basophils in particular) scatter very considerable quantities of both UV and 488-nm light 90° owing to reflection and refraction in the subcellular granules. This is picked up by the light collection system and most is filtered out before reaching the detector. However, it is never possible to remove all unwanted photons. Again, if the specific fluorescence being measured is bright there is no problem, but if it is very weak and high photomultiplier high-tension voltages are being used then a considerable proportion of the signal can be due to scattered light. This adds a source of variation similar to autofluorescence.

6.5.3   Non-specific binding

This factor is most frequently encountered in immunofluorescence work and can arise from three main sources. The first is bad preparation, where insufficient or inefficient washing steps have been carried out. The second is use of inappropriate antibodies for the system being studied, resulting in binding of antibodies to epitopes which do not belong to the test system. Finally, non-specific trapping of fluorochrome-coupled second reagents can occur, particularly when intracellular antigens are being probed.

6.5.4   Inherent variation

Inherent variation in the biology generally makes the biggest single contribution to the overall variation in the data. This is particularly true in immunofluorescence, calcium and enzyme-activity studies. It is the biological variation to which we should direct our attention, but this can be studied realistically only if all the other sources of variation contributing to the distributed response from the population are minimised.

All of the various factors summarised in this chapter can, to some extent, give rise to variation and a distributed response. This has one very important implication which does not appear to be generally appreciated. When, for example, the distributions of two overlapping immunofluorescence populations are present in the same sample, it is not only perfectly possible, but also inevitable, that there will be specifically labelled cells of low intensity which elicit a lower response from the instrument than do some non-labelled cells. *This is a direct consequence of the simple fact that the data are distributed.* This same phenomenon is seen in the deconvolution of DNA histograms, where the early S-phase cells overlap some of those in the G1 distribution. In the early days of DNA histogram analysis this was an article of faith derived from logical deduction, and absolute confirmation was achieved with the advent of the bromodeoxyuridine technique.

7

Immunofluorescence data

We are now in reasonably good shape to start analysing our data using the various criteria and procedures considered in the previous chapters. In fact, a considerable quantity of such analytical work has been carried out in the analysis of DNA histograms, as will be seen in the next chapter. In contrast, very little comparable analytical work has been performed for immunofluorescence data, but this is becoming increasingly necessary in order to resolve populations in close proximity.

The discrepancy in the quantity of analytical work carried out in these two major areas of flow cytometry (FCM) stems from the backgrounds of the people involved in those areas. The DNA histogram–analysis fraternity has hailed from cell kinetics and mathematical modelling, and these people are *au fait* with deconvoluting overlapping histograms. The immunofluorescence people tend to have purely biological backgrounds with little mathematical or statistical insight. As always, with any sweeping generalization there are exceptions; however, I'm sure the majority of immunologists (which in this context is more than 50%) would agree with me.

The difference has been one of overall approach. The DNA histogram people have been compelled to deconvolute histograms because the biology "overlaps" at the G1/S and S/G2+M interfaces. In contrast the immunologists, very worthily, have sought to improve their reagents and techniques in order to gain an increase in the signal-to-noise ratio in an effort to more efficiently pull apart populations in close proximity. I suspect, however, that there is a limit to this approach. With increasing instrument sensitivity we will find more and more subpopulations with "weak" labelling (not many epitopes per cell) which were previously thought to be unlabelled (no epitopes at all). Data sets such as these, which to some extent overlap the control sample, will always require deconvolution. Moreover, it is these populations that are likely to be most interesting, as they may signify early and subtle changes in a biological system.

7.1   Representation of the average

It is not only desirable but also necessary to represent immunofluorescence data by some "average" value for the whole histogram such as the mean,

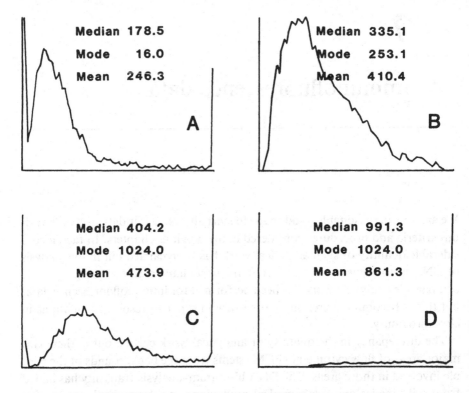

Figure 7.1. Examples of flow cytometric distributions for various cell-surface determinant assays, where the mean, median and mode are recorded on each individual panel.

median or mode. These types of data sets frequently have a wide coefficient of variation (CV) and are skewed to the right; the relationships between the mean, median and mode were illustrated in Figure 2.2. It is particularly important to choose the correct parameter to describe these types of data sets, where some cells may exceed the dynamic range of the analogue-to-digital conversion step and be scored "off scale" and appear in the first or last channel of the histogram. The problem is that we do not know how far off the scale's end these cells may be; thus, the true mean cannot be calculated. Figure 7.1 shows a number of flow cytometric distributions for various cell-surface determinant assays in which all cells should have been labelled and where the mean, median and mode have been calculated and displayed on each individual panel. The potential problems are seen in panels A, C and D. In panel A some cells were scored with either zero or very little fluorescence and the mode was calculated as 16. In panels C and D there are very bright cells which are off scale at the right and in each distribution the mode is 1024, which is meaningless. Also, the means of these distributions are underestimated, as it is not possible to know how far off scale the off-scale cells would have been recorded if this had been possible. The only measure of central tendency for the data shown in panels A, C and D is the median. As soon as more than 50% of the population is off scale then there is no measurement of central tendency that can be used.

7.2 Conventional methods

The time-honoured method for analysis of 1-dimensional immunofluorescence data has been to run a control sample, place a delimiter "by eye" at the upper boundary of the control and subsequently to score as positive all test-sample cells to the right of this delimiter. If the two distributions being compared are completely separated then there is obviously no problem with this approach. However, if there is overlap between two populations (which is more frequent than seems to be generally appreciated by commercial organizations), this method is not only hopelessly inadequate but also potentially dangerous – particularly in the clinical context, where results may influence therapeutic decisions. We will see later that this procedure can produce errors of up to one order of magnitude under some conditions.

The analytical methods described in this chapter were developed to estimate the proportions of cells in two populations embedded within an overlapping histogram of the type generated by flow cytometric immunofluorescence analysis where the labelling is weak. Both methods allow for subsequent objective statistical assessment of the validity of the estimates using Kolmogorov–Smirnov statistics, Students t and $\sum \chi^2$, so they include a majority of the "stuff" covered in previous chapters. The first procedure was designed for data sets obtained with linear amplifiers (constant CV analysis) and the second was designed for data with constant variance as obtained with log amplifiers.

7.3 Kolmogorov–Smirnov (K–S) assessment

As a prelude to both analytical procedures, the K–S statistic (described in detail in Section 4.2.2) is applied to test if the control and labelled samples are different. The two data sets used here for illustrative purposes were chosen for a number of reasons. First, in both it was considered impossible to place a vertical delimiter by eye with any degree of reliability. Second, the linear amplifier-generated data have some local historical interest. They were the first produced in Cambridge with a flow cytometer (in 1976) using Milstein's IgM monoclonal antibody (then designated Ig5.C3(H6/31)), which was specifically directed towards "IgD-like" molecules (Pearson et al. 1977). Third, the constant-variance data obtained with log amplifiers were selected as the control histogram conformed very closely to a normal distribution, with skew and kurtosis (see Section 3.1.5) of -0.18 and 2.86, respectively.

These two data sets are shown together in Figure 7.2, with the linear- and log-amplifier data (respectively) in panels A and B, and where the control data

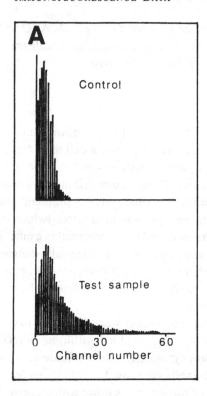 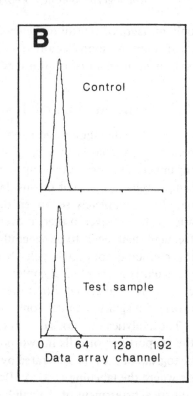

Figure 7.2. Panels A and B show immunofluorescence data obtained with linear and log amplifiers, respectively. In both panels, the control and test-sample data are displayed as the top and bottom histograms, respectively. The scale units on the abscissa of panel B are styled "data array channel", where 64, 128 and 192 correspond to 10^1, 10^2 and 10^3 respectively.

sets were obtained with an irrelevant antibody prior to staining with the fluorescenated second antibody. The control and test samples are marked on each panel for clarity. The scale units on the abscissa of panel B are styled "data array channel", where 64, 128 and 192 correspond to 10^1, 10^2 and 10^3 respectively.

The respective cumulative frequency distributions of these histograms are shown in panels A and B of Figure 7.3, where the maximum difference between control and test sample, both normalized to unity, was 0.3813 at channel 8 in panel A. There were 9070 and 10,229 cells in the control and test sample, respectively, and the difference was associated with $p < 0.001$. A comparable analysis for the panel B data gave a maximum difference of 0.0655 at channel 37 with 8976 and 8570 cells in the control and test sample, respectively; this difference was associated with $p < 0.05$. However, although both test distributions would be regarded as significantly different from their controls, the analysis does not give us a quantitation of the proportions that constitute the labelled fractions, which is what we want to know.

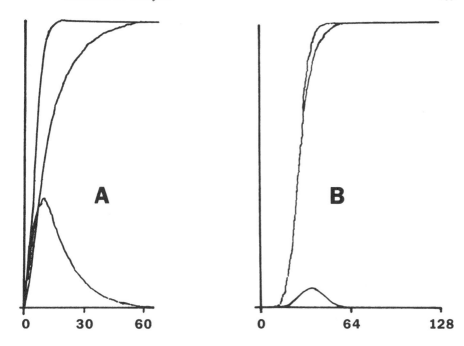

Figure 7.3. Cumulative frequency distributions of the data shown in Figure 7.2. The histogram in each panel shows the difference between the control and test-sample data.

7.4   Constant CV analysis

In this section we consider a method of analysis for deconvoluting histograms with two overlapping distributions obtained with a linear amplifier to estimate the proportions of cells in the two populations.

7.4.1   Skewed-normal distribution

We considered in Section 6.4.5 the effect of the analogue-to-digital conversion (ADC) step on the shape of a distribution produced with linear amplification. The CV of such a distribution is constant and as a consequence the ADC will produce a positive skew in the data. If the data were in fact normally distributed then they would end up looking a little like a log-normal distribution owing to the positive skewing effect. What we see is the data output from the instrument, and we have to ask the question "What is the real biological distribution from which the output data were obtained?" This question must be addressed before we can start fitting a distribution to the data.

Schuette et al. (1983) considered this problem in resolving closely spaced DNA histogram peaks with low CVs (<0.05) in delineating aneuploid tumour components close to a diploid peak. Their method used an iterative computational procedure, but for CV values of 0.04 they required up to 1600 passes through the data to obtain satisfactory resolution. With CVs in the region of 0.2, which is typical for immunofluorescence, the number of passes through the data would increase astronomically, rendering a purely iterative technique totally impracticable. The approach in our laboratory for the higher CVs seen with surface-marker data has been to use a semi-analytic approach, where systematic corrections are applied to the experimental data assuming Gaussian distributions (Watson and Walport 1985). The various steps in the development of this approach are as follows.

(1) Assume a Gaussian distribution.
(2) Calculate the distribution for given mean and SD.
(3) Transform this to a distribution with constant CV as would be generated by the flow cytometer. This requires that each channel of the true distribution be redistributed according to the CV of that particular channel. The new distribution (defined as *skewed-normal*) is then obtained by a channel-by-channel summation of the redistributed data. In practice we limited the spreading to ±4 SD of a given channel, although for some very extreme examples we encountered some computational rounding-down problems as predicted by Schuette et al. (1983).
(4) Compute the mean of the skewed-normal distribution, which should be identical to the original as an equal number of cells should be redistributed either side of the mean (although the mode and the median will differ from the original). This estimated mean is satisfactory so long as only a small proportion of the total is redistributed to the first and last channels of the finite range into which the data must be confined in order to simulate the instrument's finite range. In extreme examples a systematic error was noted for which corrections could be applied.
(5) Calculate the variance and hence CV of the skewed-normal distribution.
(6) Apply systematic corrections for the CV found at step (5) and if necessary for the mean found at step (4), and generate the reconstructed normal distribution from these values.
(7) Take the reconstructed distribution and transform it according to the procedure at step (3) and compare it with the initial skewed-normal distribution generated at step (3).

Calculation of the variance and hence the CV of the skewed-normal distributions was subject to error due to off-scale data being accumulated in the first or last channel of the finite range (1 to 256). In Figure 7.4 the first predicted CV of the skewed-normal distributions is shown versus the true CV for distributions with means of 25, 50, 100, 150 and 200 confined within a scale of 256. The limiting slope was found to be $\sqrt{2}$ when all of the skewed-normal distribution was within the 1–256 range (i.e., means of 25 and 50 with CVs less than 0.4). Thereafter, there was a systematic deviation from the $\sqrt{2}$ relationship dependent on both CV and mean. Figure 7.5 shows the first predicted CV plotted

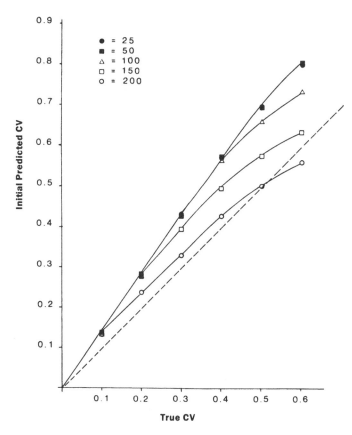

Figure 7.4. Initial predicted CV of the skewed-normal distributions versus the true CV for distributions with means of 25, 50, 100, 150 and 200 confined within a scale of 1 to 256.

against mean. There was a fairly well-defined break point at each CV level which was described by the equation

$$CV_e = -(1.15/190) \times m + 1.15,$$

where CV_e is the first estimate of the CV and m is the mean of the distribution. This equation is depicted as the dotted line in Figure 7.5. There was little deviation from the $\sqrt{2}$ relationship to the left of this line except at true CV > 0.5, but to the right there was increasing deviation proportional to both mean and CV. However, each of the extrapolations to the right passed through an abscissa scale reading of about 525 (see diagram). Because we know the mean the corrections for this domain become trivial, and minor correction for CVs above 0.5 can be included.

There was also a small systematic error in the mean calculated at step (4) with values greater than about 120 when the CV was greater than 0.15. As mentioned previously this was due to off-scale data, but this was corrected with the relationship

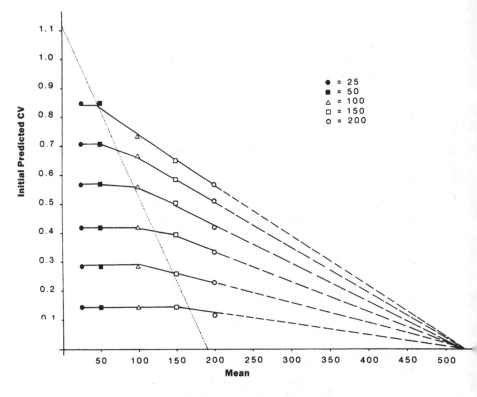

Figure 7.5. Initial predicted CV plotted against mean.

$$X_t = X_e + (((0.86 \times X_e) - 100.6) \times (CV_t - 0.15))$$

empirically derived from analysis of the data in Figure 7.5, where X_t and X_e were (respectively) the true and first estimate of the mean and where CV_t was the second estimate of the CV (cf. the previous calculation for CV_e). This correction was applied only if the mean was above 120 and the CV was greater than 0.15.

Figure 7.6 shows a matrix of histograms with means of 25, 50, 100, 150 and 200 arrayed horizontally with CVs of 0.2, 0.4 and 0.6 arrayed vertically. There

Figure 7.6 (facing page). Matrix of histograms with means of 25, 50, 100, 150 and 200 arrayed horizontally and CVs of 0.2, 0.4 and 0.6 arrayed vertically. In each panel, two histograms are bounded by dots which represent the initial histograms generated at step (2) and the regenerated histograms from step (6) (see text). The second set of histograms are bounded by the uninterrupted lines and by square dots. The latter tend to become confluent and give the impression of a continuous thick line in all panels except those towards the upper left. The histograms bounded by the uninterrupted lines are the original skewed-normal distributions, and those bounded by the square dots were computed from the reconstructed normal distributions.

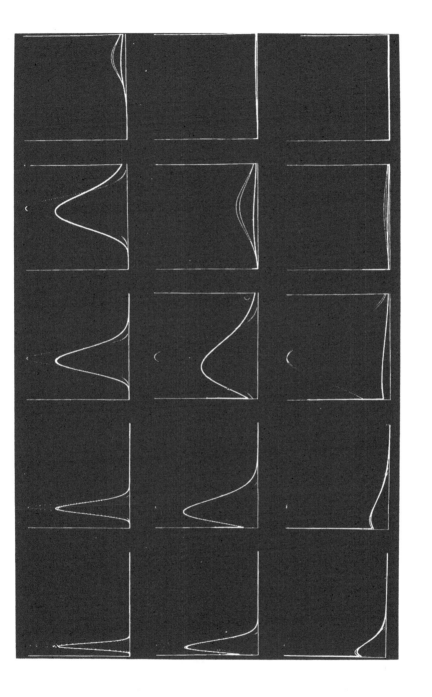

are four histograms on each panel. Two of these are bounded by the dots, which represent the initial histogram generated at step (2) and the regenerated histogram from step (6). These should overlie each other exactly, but due to computational rounding down there are some small deviations at high CVs and/or high means. The second set of histograms are bounded by the uninterrupted line and by square dots. The latter tend to become confluent and give the impression of a continuous thick line in all panels except those towards the upper left. The histogram bounded by the uninterrupted line is the original skewed-normal distribution, and that bounded by the square dots was computed from the reconstructed normal distribution. The differences between these are insignificant. Figure 7.7 shows that the corrected CV values correspond accurately with the true values.

Figure 7.8 shows an experimental data set obtained from unstained red blood cells by measuring 488-nm light scatter (argon laser) breaking through the green fluorescence photomultiplier filters at high photomultiplier voltage. This data

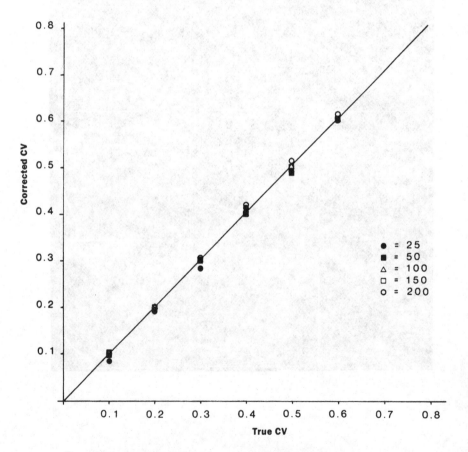

Figure 7.7. Corrected CV values versus the true values.

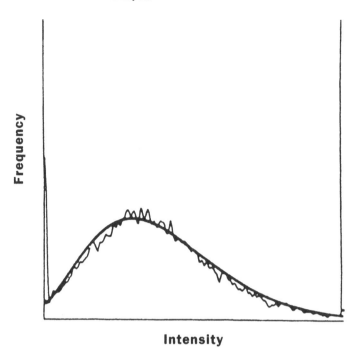

Frequency

Intensity

Figure 7.8. 488-nm scattered light from unstained red blood cells obtained by measuring the green fluorescence photomultiplier filter breakthrough at high photomultiplier voltage.

set was chosen to test the analysis, as normally distributed data with a high CV were expected. The somewhat discontinuous line bounds the experimental histogram of frequency versus light intensity on the abscissa. The smooth curve bounds the skewed-normal distribution which was generated from the reconstructed normal distribution. The mean of the latter was 99 with a CV of 0.39. The $\Sigma\chi^2$ calculated by comparing the experimental data with the predicted skewed-normal distribution was 199 with 248 degrees of freedom. This is associated with a probability $p < 0.012$ that the observed differences could have arisen by chance. Thus, we may conclude that the biological information was normally distributed and that the skew was instrumental in origin due to the constant CV derived from the ADC conversion. In the distribution shown in Figure 7.9 (uninterrupted line) the analysis failed to substantiate the hypothesis that the data were normally distributed. This data set is a light-scatter profile from an exponentially growing tumour cell line stained for DNA using a nuclear isolation technique. Scatter was measured on a second detector and the histogram contains data from G1, S and G2 cells; we would not expect these data to be normally distributed, as the population was far from homogeneous.

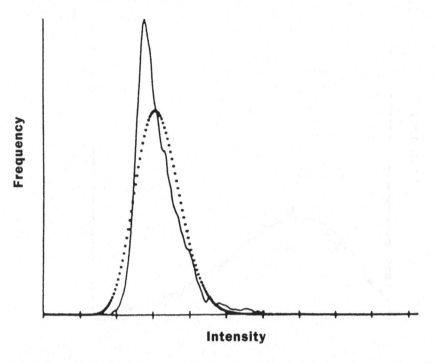

Figure 7.9. Light-scatter profile from an exponentially growing tumour cell line stained for DNA using a nuclear isolation technique.

7.4.2   Histogram deconvolution

The control data of Figure 7.2A were analysed using the above technique, which predicted a mean and standard deviation of 5.27 and 2.1 respectively. $\sum \chi^2$ was 10.5 for the observed differences between the data and the skewed-normal distribution, with 14 degrees of freedom. Entering the χ^2 table with this number of degrees of freedom, we find a value of 26 at the 0.05 probability level and 32 at the 0.01 probability level. Hence, these results are compatible with the hypothesis that the data were derived from a normal distribution.

We now assume that any unlabelled cells in the test sample of Figure 7.2A are also normally distributed, and that these can be described with $\bar{x} = 5.27$ and $s = 2.1$. A first, very approximate, value for the proportion of unlabelled cells in the test sample was obtained by doubling the area to the left of this mean value in the test sample and comparing this with the whole area. The left side of the mean was used, as this is the area of the test sample least likely to have overlap of labelled cells. A normal distribution was then generated with $\bar{x} = 5.27$ and $s = 2.1$ containing this proportion, and the distribution was skewed as in step (3) of Section 7.4.1. The resulting skewed-normal distribution was then subtracted from the test data set on a channel-by-channel basis. There were no negative numbers in the channels of the test sample from which the subtractions were carried out, hence this initial procedure underestimated the unlabelled-cell fraction (UF) which was taken as the minimum of the iterative range. The maximum in the range was obtained by incrementing successively the initial estimate of the UF by 10% until negative values were obtained. If the initial UF estimate had produced negative values after the subtraction then this would have been taken as the maximum, and the minimum would have been found by successive subtraction of 10%.

The analysis continued by using initial increments of 0.05 within the estimated range of 20% for the unlabelled population, thus five values were considered. A skewed-normal distribution of the unlabelled fraction was generated for each of these values which was then subtracted, on a channel-by-channel basis, from the test distribution. The remaining distribution was then analysed using the skewed-normal technique to give a mean and standard deviation of the predicted labelled fraction. A theoretical normal distribution for the latter was generated containing $(1.0 - UF) \times N$ cells, where N represents the number of cells in the test distribution. This was now skewed and then added, again on a channel-by-channel basis, to the skewed-normal distribution of the unlabelled fraction. This theoretical distribution was now compared with the experimental test sample data, and $\sum \chi^2$ for observed (experimental) versus expected (theoretical) was calculated. The value for the unlabelled fraction associated with the minimum $\sum \chi^2$ was now used as the mid-point of the final range, which was taken to be ± 0.05 of this mid-point. The increment was reduced to 0.01 and the analysis was repeated, which yielded an estimate of the unlabelled fraction in the test sample with a resolution of $\pm 1\%$.

The final result of the analysis is shown in Figure 7.10, where the symbols represent the experimental data and the dashed curves represent the labelled and unlabelled fractions. The continuous curve passing through the experimental data points is the theroetical reconstruction of the experimental data, which gave a total $\sum \chi^2$ of 42.7 with 60 degrees of freedom. This is associated with $p > 0.95$, meaning that the fit was good. The labelled fraction was predicted to contain 5287 cells (52%) with a mean in channel 17.2 and standard deviation of 6.92. We are now in a position to assign a significance level to this result using Student's t, where $n_1 = 4942$, $\bar{x}_1 = 5.27$ and $s_1 = 2.1$ are the parameters for the unlabelled fraction and where $n_2 = 5287$, $\bar{x}_2 = 17.2$ and $s_2 = 6.92$ are the parameters for the labelled fraction. Thus, the standard error of difference is given by

$$ SE = \sqrt{\left(\frac{4942 \times 2.10^2 + 5287 \times 6.92^2}{4942 + 5287 - 2} \right) \times \left(\frac{1}{4942} + \frac{1}{5287} \right)} = 0.1026, $$

which gives $t = (17.2 - 5.27)/0.1026 = 116$ with 10,227 degrees of freedom, a highly significant result ($p < 0.0001$). Thus, with these numbers of cells involved, we can be more than 99.99% certain that there were two populations in the test sample with 52% being labelled.

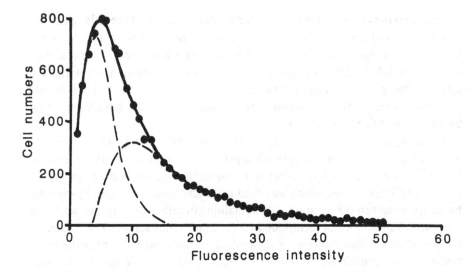

Figure 7.10. Result of the analysis of the data in Figure 7.2A, where the closed circles represent the experimental data. The dashed curves represent the theoretical reconstruction of the labelled and unlabelled fractions.

7.5   Constant-variance analysis

A different approach is used with constant-variance data. This approach does not have the added complication of ADC skewing, as the data were obtained using log amplification. Hence, the mean and standard deviation of a distribution can be obtained directly without having to use the skew analysis technique. The mean of the control in panel B of Figure 7.2, $\bar{x}_1$, was 29.1 with a standard deviation of 6.69. The mean of the test sample in Figure 7.2B was 30.5, a shift of only 1.4 linear "data array–channel" units, but the standard deviation cannot meaningfully be calculated since this is not a single distribution (as revealed by K–S analysis in Figure 7.3B). Linear units were used on the abscissae of Figures 7.2B and 7.3B, because the ensuing calculations must be performed in these linear units.

7.5.1   Ratio analysis of means (RAM)

This aspect of the analysis is derived from simple mechanics, where moments are taken about a point (Watson in press). Consider the weightless beam depicted in Figure 7.11, where masses W_1 and W_2 are suspended from the beam at points $\bar{x}_1$ and $\bar{x}_2$, respectively. If the origin of the scale (zero) is set at the extreme left of the beam then the balance point B can be obtained by taking moments, as in the following relationship:

$$W_1(B - \bar{x}_1) = W_2(\bar{x}_2 - B). \tag{7.1}$$

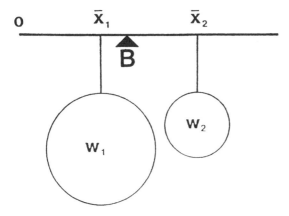

Figure 7.11. Depiction of a weightless beam where two weights, W_1 and W_2, are suspended from points $\bar{x}_1$ and $\bar{x}_2$, respectively. B is the balance point when these weights have a mass ratio of $0.75:0.25$.

On rearrangement we obtain

$$B = (W_1\bar{x}_1 + W_2\bar{x}_2)/(W_1 + W_2). \tag{7.2}$$

As a simple example, let us suppose that $\bar{x}_1$, $\bar{x}_2$, W_1 and W_2 have values of 100, 200, 75 and 25 respectively. We can now calculate B from equation (7.2) to be 125. Thus, the balance point (as depicted in Figure 7.11) is 25 distance units from $\bar{x}_1$ and 75 distance units from $\bar{x}_2$.

It should be noted that neither the shapes nor the densities of the masses depicted as hanging from the beam in Figure 7.11 (nor, indeed, their orientations) are of any consequence, as long as the forces through the centers of gravity of the masses pass through the points $\bar{x}_1$ and $\bar{x}_2$. This is illustrated in Figure 7.12, where the masses are depicted as normal distributions with the same variance cut from a material of uniform thickness. Although these distributions are overlapping, the beam will still balance at the same point because their areas have a $75:25$ ratio and the material is uniform. If you are not convinced by this argument then turn (in your mind's eye) the two distributions, which are flat, through 90° so that they are no longer overlapping.

Let us now suppose that the distances $\bar{x}_1$, $\bar{x}_2$ and B are known for a normalised total mass of unity, where $W_1 + W_2 = 1.0$. We can now calculate the relative proportion of W_2 from equation (7.1) by replacing W_1 with $(1.0 - W_2)$. This gives

$$(1.0 - W_2)(B - \bar{x}_1) = W_2(\bar{x}_2 - B),$$

which on simplification reduces to

$$W_2 = (B - \bar{x}_1)/(\bar{x}_2 - \bar{x}_1). \tag{7.3}$$

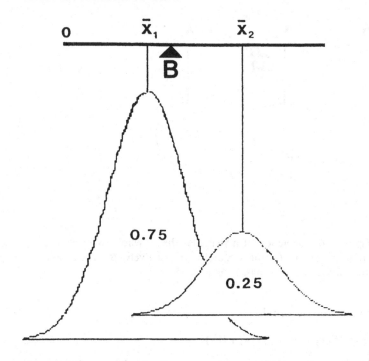

Figure 7.12. The same beam as shown in Figure 7.11, where the weights have been replaced with masses in the shape of normal distributions cut from a uniform material. The beam still balances at the same point *B* in spite of the overlap between the distributions.

The masses in Figure 7.12 were depicted as normal distributions in order to illustrate the connection between the concept of taking moments about a point in mechanics and immunofluorescence histogram analysis. We can see from equation (7.3) that the proportion W_2, which will now be referred to as labelled cells in a mixed population of labelled plus unlabelled, is defined by three "distances" – namely, $\bar{x}_1$, $\bar{x}_2$ and B. $\bar{x}_1$ is the mean of the control unlabelled fraction, B is the mean of the test sample that contains both labelled and unlabelled cells and $\bar{x}_2$ is the mean of the labelled-cell fraction. Values for $\bar{x}_1$ and B can be obtained directly from the experimental data but we also require a value for $\bar{x}_2$, which is calculated as described in the next section.

7.5.2   Labelled-fraction mean calculation

This component of the analysis arose from a desire to be absolutely sure that the K–S statistic for analysing differences between distributions was being correctly implemented within our software package. Consider the top panel of Figure 7.13, which shows two normal distributions, where both have standard deviations of 16 and means in channels 64 ($\bar{x}_1$) and 192 ($\bar{x}_2$) respectively. In order to test that the correct K–S difference was being computed,

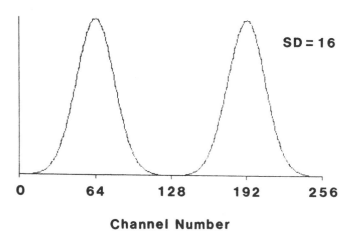

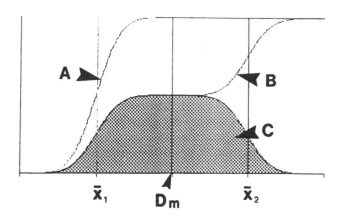

Figure 7.13. Upper panel shows two normal distributions both with standard deviation of 16 channels with respective means, $\bar{x}_1$ and $\bar{x}_2$, in channels 64 and 192. Lower panel shows the cumulative frequency distributions of the histogram with mean at $\bar{x}_1$ as the curve labelled A and that of both histograms combined as curve B. The stippled area histogram, C, was obtained by subtracting curve B from curve A. D_m is the mean of the cumulative frequency subtraction histogram. With no correction for class interval (CI), this is positioned 0.5 channels to the left of the mid-point between $\bar{x}_1$ and $\bar{x}_2$ (see text).

cumulative frequencies of the left histogram and of both histograms combined were displayed. These are shown in the lower panel as curves A and B, respectively, where both are normalised to unity. A difference histogram was then generated by subtracting curve B from curve A, which is shown as the stippled area C. It was now noted that the mean of this difference histogram, D_m (middle vertical line), appeared to be very close to the mid-point of the means (flanking vertical lines) of the histograms. In fact, as can be seen in this figure, D_m is just to the left of the mid-point between the histogram means.

This observation is potentially important because, if confirmed, it would enable the exact position of the mean $\bar{x}_2$ of the second distribution to be calculated as

$$\bar{x}_2 = (2.0 \times D_m) - \bar{x}_1. \tag{7.4}$$

Hence, an approximate value for W_2 would be obtainable via equation (7.3).

A large series of simulations was carried out for varying standard deviations, ratios of means and relative proportions in control and test samples, where cumulative frequency subtraction of the two distributions was carried out. In each case D_m was calculated to be exactly 0.5 channels to the left of the mid-point between $\bar{x}_1$ and $\bar{x}_2$. This relates to the fact that the distributions being considered are not strictly continuous, as they have a class interval (abscissa scale units) of unity. However, in the limiting case, where the distributions are absolutely continuous the class interval is infinitely small, and equation (7.4) should hold exactly. However, the latter may be corrected for class interval (CI) where:

$$\bar{x}_1 = (2.0 \times (D_m + (CI/2.0))) - \bar{x}_1. \tag{7.5}$$

A formal mathematical proof of the finding depicted by equation (7.5) is given in Appendix 8.

Figure 7.14 shows four simulations where the class-interval correction has been included. The standard deviation was 16 in each distribution and the unlabelled (a, mean $\bar{x}_1$ in channel 64) and labelled (b) fractions each constitute 50% of the total. The histograms are shown in the left column with the corresponding cumulative frequency distributions in the right column. Ratios of the means ($\bar{x}_2/\bar{x}_1$ or R_m) are shown between each pair of displays; these were set at 3.0, 2.0, 1.5 and 1.25 respectively. Note that as the labelled fraction b becomes closer to the unlabelled fraction a, there is a decrease in the maximum frequency of the subtracted cumulative frequency distribution; however, this occurs only when the distributions a and b overlap. Nevertheless, the relationship between $\bar{x}_1$, D_m and $\bar{x}_2$ is maintained.

The solution for $\bar{x}_2$ shown in equation (7.5) can now be substituted into equation (7.3), which on simplification for data with a class interval of unity gives

$$W_2 = (B - \bar{x}_1)/(2.0 \times (D_m + 0.5 - \bar{x}_1)). \tag{7.6}$$

This equation gives the proportion W_2 of labelled cells in a mixed population of labelled plus unlabelled in terms of $\bar{x}_1$, the mean of the control sample; B, the mean of the test sample containing labelled plus unlabelled cells; and D_m, the mean of the subtracted cumulative frequency distribution. It is important to note that equation (7.6) should be applied only if the variances of the control and labelled cell distributions are equal. This condition is most likely met in FCM when log amplifiers are used, as the standard deviation is independent

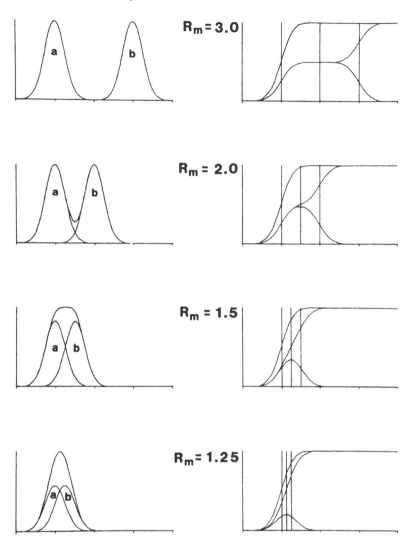

Figure 7.14. A series of four sets of identical histograms, a and b, both with standard deviation of 16 channels, where the mean of histogram a is fixed in channel 64 and where the ratio R_m of $\bar{x}_2/\bar{x}_1$ is varied from 3.0, 2.0, 1.5 and 1.25 as shown. The panels in the right column show the cumulative frequency distributions of a and of a + b combined together with the respective cumulative frequency subtraction histograms, where D_m has been corrected for class interval.

of intensity. Equation (7.6) cannot be applied directly to data obtained with linear amplifiers, where the coefficient of variation is constant.

Figure 7.15 shows the results of the cumulative frequency subtraction analysis for the control and test-sample distributions of Figure 7.2B, where $\bar{x}_1$, D_m and $\bar{x}_2$ are marked by the vertical lines on the figure together with their values, which were 29.1, 37.9 and 46.7 respectively. The calculated value of 46.7 for $\bar{x}_2$

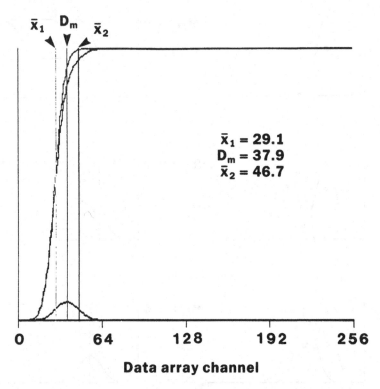

$$\bar{x}_1 = 29.1$$
$$D_m = 37.9$$
$$\bar{x}_2 = 46.7$$

Data array channel

Figure 7.15. Cumulative frequency distributions of the control and test sample shown in Figure 7.2B, with the cumulative frequency subtraction histogram displayed close to the X-axis. The latter was analysed to give D_m and hence $\bar{x}_2$ which are shown, together with $\bar{x}_1$, as vertical lines.

was now used with the value of 29.1 for the control mean $\bar{x}_1$, and the mean of 30.5 for the test sample B, to calculate the labelled fraction as 8.0%. The predicted unlabelled and labelled populations are shown in Figure 7.16 together with the experimental data.

In the first log-amplification decade (1–64) we can convert to fluorescence intensity units (FIU) with the relationship $FIU = 10^{(x/64)}$, where x is the data array–channel number. Thus 29.1, 30.5 and 46.7 translate to 2.849, 2.996 and 5.366 fluorescence intensity units respectively. Hence, in FIU the test sample mean, 2.996, is only 5.2% greater in intensity than the control, 2.849. Nevertheless, in spite of this very small shift, the analysis predicts 8.0% of the total population to be labelled.

7.5.3 Statistical verification

We now have to ask if this is reasonable and what significance can be placed on the result. This is particularly important when analysis of data predicts a minority labelled subset embedded within the test sample. There are three ways in which the predictions can be verified: K–S analysis, Student's t, and simulation of the experimental data using $\sum \chi^2$ as the criterion of fit.

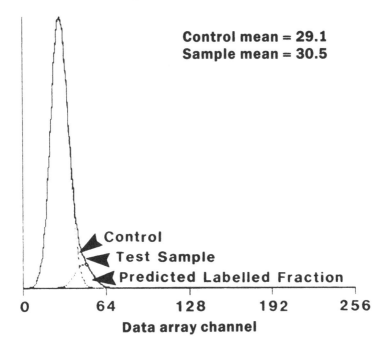

Control mean = 29.1
Sample mean = 30.5

Control
Test Sample
Predicted Labelled Fraction

0   64   128   192   256
Data array channel

Figure 7.16. A control and test-sample data set as indicated, together with the predicted labelled-fraction histogram. The means of the control and test sample were 29.1 and 30.5 respectively. Note that the abscissa scale units are expressed as "data array channels", where 64, 128, 192 and 256 correspond to 10^1, 10^2, 10^3 and 10^4 fluorescence intensity units (FIU) respectively.

Kolmogorov–Smirnov analysis

The cumulative frequency distributions of the control and test sample were re-analysed over a range of ± 3 standard deviations about the mean of the predicted labelled distribution $\bar{x}_2$. These data are shown in Figure 7.17 together with the difference histogram close to the abscissa. With the number of cells involved, K–S analysis showed that the two cumulative frequency distributions over this limited range had a probability of being different at the 99% confidence limit.

Student's t

The predicted results were also submitted to Student's t analysis, assuming the variance was equal in both distributions. The standard error of difference, SE, was calculated (as detailed in Section 4.1.2):

$$SE = \sqrt{\left(\frac{9171 \times 6.69^2 + 797 \times 6.69^2}{9171 + 797 - 2}\right) \times \left(\frac{1}{9171} + \frac{1}{797}\right)} = 0.2463.$$

The difference between means was 17.6 channels which, when divided by 0.2463 (the standard error of difference), gives $t = 71.46$ with 9966 degrees of freedom. This is significant at $p < 0.0001$.

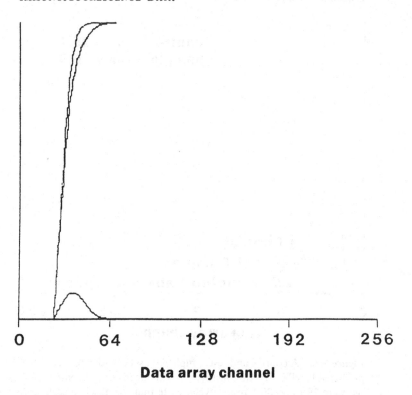

0   64   128   192   256

Data array channel

Figure 7.17. Kolmogorov–Smirnov analysis of the control and test-sample distributions over the interval $\bar{x}_2 \pm 3$ SD. The difference histogram is shown close to the X-axis, and the maximum height of this is the maximum difference between the two cumulative frequency distributions, which was significant at $p > 0.99$.

Simulation with $\sum \chi^2$

A final verification of the analysis procedure was carried out by generating two normal distributions with $\bar{x}_1$ and $\bar{x}_2$ of 29.1 and 46.7, respectively, both with a standard deviation of 6.69, where the proportions in the two distributions were 0.92 and 0.08. These distributions were summed on a channel-by-channel basis, and the synthetic data set was normalised to a total area of 9968 to correspond to the number of cells in the test sample data set. A $\sum \chi^2$ calculation was then used as the criterion for the difference between the data and the histogram synthesized from the results of the data analysis. This gave a probability of $p < 0.0001$ that the difference between the simulated and experimental data could have arisen by chance.

7.6   Errors with conventional methods

A further analysis of 95 synthetic histograms was carried out to tabulate the errors incurred with conventional analysis, where the labelled fraction

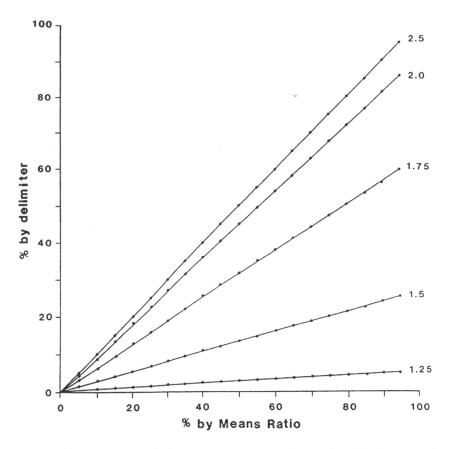

Figure 7.18. Labelled percentages calculated by vertical delimiter plotted on the ordinate versus labelled percentages calculated by RAM on the abscissa for means ratios of 2.5, 2.0, 1.75, 1.5 and 1.25.

is calculated using a vertical delimiter set at 2.5 standard deviations above the mean of the control distribution. The mean of the latter was set in channel 32 and both the control and labelled fraction distributions were synthesised with a standard deviation of 8 channels (CV = 0.25 in channel 32). The labelled fraction was then varied from 5% to 95% in increments of 5% for ratios of labelled to unlabelled fraction means of 2.5, 2.0, 1.75, 1.5 and 1.25. These results are shown in Figure 7.18, where the labelled percentages calculated by vertical delimiter are plotted on the ordinate versus labelled percentages calculated by this method (ratio analysis of means) plotted on the abscissa. As long as the data are normally distributed and the variance is constant, the error calculated by RAM is less than 0.5%. We can see from Figure 7.18 that when the means ratio was 2.5 the error incurred by using the vertical delimiter was zero, as the channel separation between the two means was 43 (75 − 32). This is greater than five standard deviations (5 × 8), 2.5 above the lower mean and 2.5 below the upper mean. However, as the separation between the means decreased the errors increased dramatically, and with a separation of about 1.3 the error to be expected reaches approximtely one order of magnitude.

7.7   Concluding remarks

The methods described in this chapter were undertaken to help in the analysis of immunofluorescence data by presenting results in the format alluded to in the introductory comments of Chapter 4. For the first data set we can say that the analysis was compatible with two populations, where the respective means were in channels 5.27 and 17.2 and the labelled fraction comprised 52% of the total. Moreover, the separation between the means (with the standard deviations involved, 2.1 and 6.92 respectively) could have occurred in less than 1 trial in 10,000 on a random basis.

In the second data set the analysis was also compatible with two populations whose means were separated by 1.6 data-array channels, where the predicted positive fraction constituted 8.0% of the total at or above the 99.9% confidence interval. If we had just "eye-balled" the data in Figure 7.2 and placed a vertical delimiter, we could not possibly have come to the conclusion that 52% and 8.0% of the respective populations were labelled. Yet our objective statistical analysis tells us there is a probability exceeding 0.999 that there were two populations in each sample. One of the reasons we can have this degree of certainty is that a large number of cells was analysed, which is one of the frequently voiced advantages of the technology. However, the large-number "good-statistics" attribute of the technology is very infrequently used to its full advantage.

The objective statistical assessments described in this chapter cover a comparison of the experimentally derived control and test-sample data using Kolmogorov–Smirnov statistics, Student's t assessment of the differences between the predicted labelled and unlabelled distributions and a $\sum \chi^2$ to test for differences between the predicted and experimental data. In the constant-variance analysis this represents a three-way comparison – namely experimental versus experimental, theoretical versus theoretical and experimental versus theoretical.

At present the second method of analysis can be used only for data sets with constant variance, as illustrated with distributions where both skew and kurtosis conformed very well with a Gaussian profile (indeed, these data were chosen for this property). However, the technique can also be applied to data that depart from normality as long as the skew is close to zero and the variance is constant. At the extreme, for example, if the distributions being analysed have rectangular profiles then the cumulative frequency curves will be oblique straight lines. This will result in a rhomboid or triangular cumulative frequency subtraction histogram, but with constant variance the shape will be symmetrical and the slopes of the lateral two oblique sides will be equal and opposite. Hence, the mean of the subtraction histogram will still be midway between the means of the component distributions, subject to class-interval correction.

In general, the distribution of biological behaviour tends to be skewed to the right and of log-normal type, and any analogue-to-digital conversion step in combination with linear amplification introduces a further positive skew (Schuette 1983; Watson and Walport 1985). With log amplification the positive skewing tends to be eliminated and log-normal distributions present themselves in Gaussian form, which is highly convenient for the constant-variance analysis described here. Furthermore, inspection of many published immunofluorescence data sets suggests that the variance is not infrequently constant, as the shapes of well-separated control and test-sample distributions tend to be very similar. However, with the overlapping data for which the constant-variance method is intended, it is not possible directly to test if the variance of the unlabelled and labelled fractions is constant. It is likely that this will be so with log amplification, at least to within reasonable limits, but this is an assumption and not an axiom. The final phase of this analysis attempts to address this point by reconstructing the test-sample distribution from the results of the analysis, where an objective method of assessing the difference between experimental and reconstructed data using $\Sigma \chi^2$ is used. This is important in order to identify discrepancies that are most likely to be due to non-constant variance, a problem which, from initial studies, appears to be associated with systematic errors for which corrections should be able to be made.

8

DNA histogram analysis

8.1   The cell cycle

In 1951 Howard and Pelc discovered that the DNA synthesis period occupied a discrete interval separated temporally from mitosis. The immediate conclusion was that there must be at least four distinct phases within the division cycle, one of which (mitosis) had already been defined. They termed the post-mitotic pre-synthetic phase G1 and the post-synthetic pre-mitotic phase G2. The G designation stands for "gap", and a representation of this concept of the cell cycle, which stands to this day, is shown in Figure 8.1. In 1959 Quastler and Sherman introduced the percent labelled mitosis (PLM) curve from which the durations of the intermitotic phase times could be deduced. In this technique cells are exposed to a pulse of tritiated thymidine, a radiolabelled precursor of thymine, which is incorporated into DNA only during the DNA synthesis period. At regular intervals thereafter autoradiographs are prepared, and the percentage of mitotic figures in the population which are labelled with silver grains is scored. These percentages are plotted against the time between labelling and sampling, which gives a damped sigmoidal curve whose periodicity is that of the cell cycle. Computer model analysis of the PLM curve gives an estimate of the intermitotic phase times with their variances.

Almost as soon as the technique became available it was apparent that the cell-cycle time was invariably less than that of the doubling time of tumours growing in vivo. This gave rise to the conclusion that only a proportion of cells in the tumour were within the division cycle, and Mendelsohn (1962) coined the phrase "growth fraction" for this compartment. An extra cell-cycle phase, G-zero or G_0, was proposed to describe immediately post-mitotic cells that spent some time in a resting phase before re-entering the division cycle. In some tissues there is good evidence for the existence of such a compartment (e.g. peripheral lymphocytes that can be stimulated to divide by mitogens), but in many tumour tissues the evidence is tenuous.

The PLM technique is labour-intensive and very time-consuming; it can take up to six weeks to obtain a result for the cycle time and that of the intermitotic

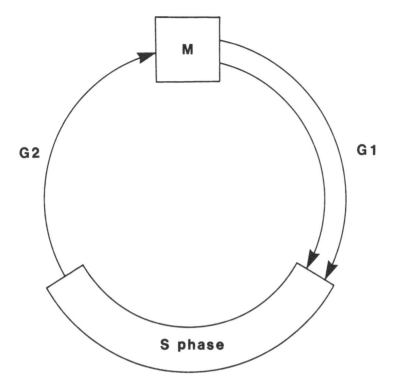

Figure 8.1. The cell cycle (after Howard and Pelc 1951).

phase times. The results obtained, namely the time it takes to complete the various phases, are generally not very interesting in themselves. However, specific biochemical manipulation of cells which induce changes in the phase times can give some insight into the biological processes modified by the manipulation. They can also be used to estimate the growth fraction in human tumours.

Flow cytometric techniques have now completely revolutionised cell-cycle kinetics. We can obtain not only the cell-cycle and intermitotic phase times but also the proportion in S phase which is actively synthetic; we can discriminate these cells from those arrested with an S-phase DNA content in little more than the cell-cycle time. In terms of hours saved this represents a gain factor of about 100-fold, quite apart from the fact that we obtain information that could not otherwise be acquired: namely, the distinction between S-phase DNA content cells that are synthesizing and those that are not. Furthermore, there are techniques for some types of cells which discriminate between G_0 and G1 and subsets within these groups.

8.2   The DNA histogram

The DNA histogram is a very simple data set that characteristically contains two peaks separated by a trough. The first peak, which is usually the

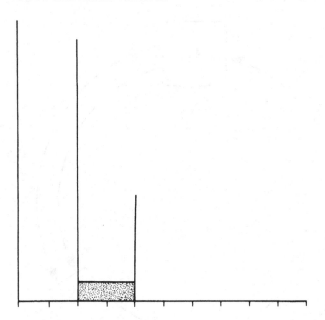

Figure 8.2. Idealised DNA histogram with no dispersion, where all G1 and G2 + M cells are recorded in channels 200 and 400 respectively.

larger, corresponds to cells with $G_0/G1$ DNA content and the second, which should be at double the fluorescence intensity of the first, corresponds to cells with G2 + M DNA content. Any cell scored in the trough has a DNA content intermediate between G1 and G2 + M, and these *usually* represent cells in S phase. In a perfect data set (which doesn't exist), all G1 and G2 + M cells would be scored in single channels and any cells between or immediately adjacent to these would be in S phase; this is shown in Figure 8.2. In practice, however, the data are distributed owing to a number of factors (to be described shortly) and the effect of this is shown in Figure 8.3, where the dispersion is assumed to be Gaussian.

I stated that cells in the trough usually represent cells in S phase, stressing the "usually". In unperturbed populations these cells are in S phase; however, in populations perturbed by therapeutic intervention, radiation or drugs, it is possible for cells to have an S-phase DNA content when they are either temporarily arrested or dead and no longer synthesising. Examples of this state will be given later, and it is obviously important to make this distinction. The best analogy I have heard to describe this comes from Professor Len Lamerton: "You can sit down at the table to have dinner but you don't have to eat anything". (He later admitted that he had heard this from someone else, but couldn't remember who!)

Dispersion in the data is due to instrumental, staining and biological factors, and potential sources of variation were considered in Chapter 6. Variability in

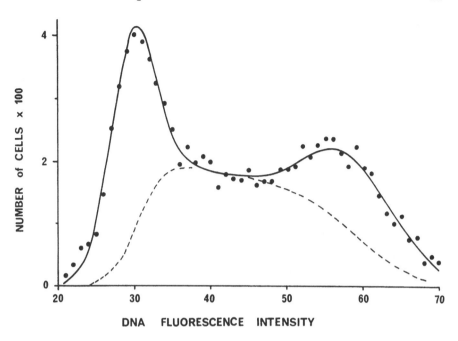

Figure 8.3. Example of a real DNA histogram with dispersion and a CV in the data of about 7%. The G1 and G2+M peaks, whose true modal positions are channels 30 and 60 respectively, are superimposed on the distributed S-phase component. This shifts the apparent G1 modal channel towards G2+M and that of G2+M towards G1, hence the G2+M:G1 ratio will appear to be less than 2.0.

illumination intensity due to core positional instability (see Section 6.2.3) is probably the major factor, but staining variability between cells is another. This is most likely to occur when heterogeneous populations are being studied. These include normal tissues where there are different degrees of differentiation of cells within the sample as well as tumour samples. We have already seen that a particular molecule, DNA in this context, is measured indirectly by fluorescence which is proportional the number of dye molecules bound to DNA, and the number of binding sites can vary with the conformation of DNA and chromatin. Not only the quantity but also the energy of fluorescence from some stains (e.g. acridine orange and Hoechst 33342) are sensitive to the number of accessible binding sites and the state of the microenvironment. There may also be variation in the amount of DNA from cell to cell within apparently homogeneous populations. The mouse mammary EMT6/M/CC cell line, originally derived by Rockwell, Kallman and Fajardo (1972), has a between-cell variation in its chromosome number from 50 to 70 (Watson 1977b). Furthermore, the binding of ligands to DNA (or indeed anything) is governed by dynamic processes. Wherever there is an association binding constant there is always a dissociation constant (see Figure 6.1), and the results that we see are governed by the law of mass action. Nicolini et al. (1979) have discussed this at length in relation to acridine orange staining. In order to think realistically about DNA histograms we must consider the number of dye molecules in relation to the number of accessible binding sites. Although the former can be controlled experimentally, the latter may not be controllable, as differences in chromatin structure in different cells and its organization in G1, S and G2+M in the same cells may make differences to the quantity of fluorochrome that can be bound per DNA phosphate residue.

The ratio of the means of the G2+M:G1 peaks should be 2.0, but an ADC offset (see Section 6.4.5) and differences in the number of available dye binding sites in G2+M and G1 may result in a ratio that is not equal to 2.0. Moreover, it is not generally appreciated that dispersion in the data can also affect this ratio. Consider Figure 8.3. The slope of the S/G2+M interface is less steep than that of the G1/S interface, and the Gaussian distributed G2+M and G1 peaks are "added on" to these respective interfaces. This results in a shift of the position of the G1 peak towards G2+M and a shift of the latter towards G1 (Watson 1977b).

8.3   DNA histogram analysis

The objective of DNA histogram analysis is to obtain a reasonable approximation for the proportions of cells in the G1, S and G2+M phases of the cell cycle. A great deal of time and effort has been expended on this using computer models of varying complexity (Baisch, Gohde and Linden 1975; Baisch and Beck 1978; Beck 1978; Christensen et al. 1978; Dean and Jett 1974; Fox 1980; Fried 1976, 1977; Fried and Mandel 1979; Gray 1974, 1976, 1980; Gray, Dean and Mendelsohn 1979; Jett 1978; Johnston, White and Barlogie 1978; Kim and Perry 1977; MacDonald 1975; Watson 1977b; Watson and Taylor 1978; Zeitz and Micolini 1978). A comparative review of these various methods has been published by Baisch et al. (1982). As expected, some models performed better than others, not only with simulated but also with experimentally derived data; however, no model was ideal. The models that performed best tended to be those requiring a large mainframe computer, and these also tended to contain a large number of variables for which a solution had to be found. It is not clear, however, whether it was the computing power available on the mainframes or rather the intrinsic structure of the models which contributed to the better performances. All models contain the common assumption that the G1 and G2+M peaks are Gaussian distributed, but they differ in the way in which the S-phase component is calculated.

8.3.1   Age distribution theory

Many models base the S-phase calculation on age distribution theory in exponentially growing populations (Steel 1968), which is summarised in Fig-

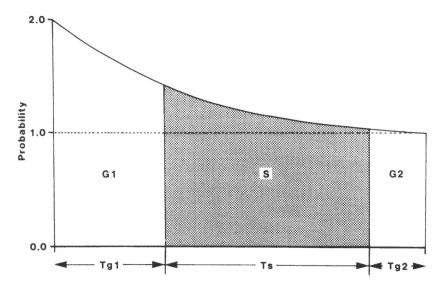

Figure 8.4. Age distribution of exponentially growing populations (after Steel 1968).

ure 8.4. In this diagram we assume, not unreasonably, that each cell divides into two at mitosis, and because of this assumption we define the probability of finding a cell of zero age (immediately after division) as 2.0. Hence, the probability of finding a cell of age unit-cycle time, t_c, is equal to unity where $t_c = t_{G1} + t_S + t_{G2+M}$ and where t_{G1}, t_S and t_{G2+M} are the durations of G1, S and G2 + M, respectively. The probability boundary falls exponentially from 2 to 1 throughout the cell cycle as the population is growing exponentially. The proportions of cells in G1, S and G2 + M are now obtained by finding the area (which is done by integration, but I won't go into that) within the age distribution occupied by the three phases, and that for S phase is shaded in Figure 8.4. When we translate the age distribution diagram into DNA histogram terms, all those cells in the G1 region have the same DNA content and should be scored in a single channel (see Figure 8.2). Similarly, all those in G2 + M should also be scored in a single channel at double the abscissa scale reading. The S-phase area of the age distribution spans the interval between the G1 and G2 + M peaks, and the shape of its upper surface will be identical to the upper S-phase boundary of the age distribution diagram (subject to scaling factors) if the rate of DNA synthesis is constant.

In practice, we have seen that the G1 and G2 + M peaks do not appear in a single channel of the DNA histogram. In order to apply age distribution theory to the analysis of DNA histograms we must compute the relative durations of the three intermitotic phase times (G1, S and G2 + M), compute the growth fraction as some proportion between zero and unity, and assume that the rate of DNA synthesis is constant. We must also remember that there is a spread in the relative phase durations which must be assumed to have some probability function (e.g. a normal distribution), which additionally requires the standard deviations of the phase times, σ_{tG1}, σ_{tS} and σ_{tG2+M}, to be considered. Thus far we have defined seven variables that may have to be computed, and if there is any suspicion that the DNA synthesis rate is not constant then a minimum of two more variables would have to be added to the computation.

Some of the kinetic variables that relate to population growth have now been considered, but there are also a number of static variables relating to the DNA histogram. These include the positions of the G1 and G2+M peaks, their standard deviations and the G2+M:G1 ratio. We have now defined a total of 14 variables that may have to be considered in the computational procedure, and it is clear that the type of data set shown in Figure 8.3 is far too simple to be able to support a model with this number of variables. The time taken for this type of computation can become prohibitive without materially altering the results obtained; all that happens is that the uncertainty in the values of the computed parameters increases. Age distribution theory as outlined here should be used only for data sets where there is likely to be a steady state of exponential growth and where the rate of DNA synthesis is constant. These conditions are frequently approximated in tissue culture when the population is not perturbed by therapeutic intervention. However, in the real world of in vivo tumours (including human) and therapeutically perturbed populations, the age distribution is distorted and can no longer be modelled by a continuous mathematical function. Because of these problems, some simpler approaches have been adopted to model the S-phase interval of the histogram.

8.3.2   Rectilinear integration

This term is a somewhat grandiose description for finding the area, in this context, of a four-sided polygon with straight edges. It would seem that when you have a sophisticated instrument you must have "flashy" terms to describe some very simple operations. Consider Figure 8.5, which shows two hypothetical DNA histograms. In panel A the S-phase section of the histogram has a horizontal upper border. The height of this above the baseline can be found and multiplied by the distance between the G2+M and G1 mean channels to give an area approximately proportional to the fraction in S-phase. In practice the height would be calculated as the average of the number of channels between points α and β. This procedure extrapolates the upper border (dashed lines) of the S-phase interval of the histogram to the mean channels of G1 and G2+M and gives the S-phase proportion.

In panel B the upper border of the S-phase interval is not horizontal but it is still straight. Thus, if points α and β are equidistant from the G1 and G2+M mean channels (respectively) then we can still compute the average and multiply this by the distance between the means to give the area of the polygon that is equivalent to the S-phase proportion. This method was introduced by Baisch

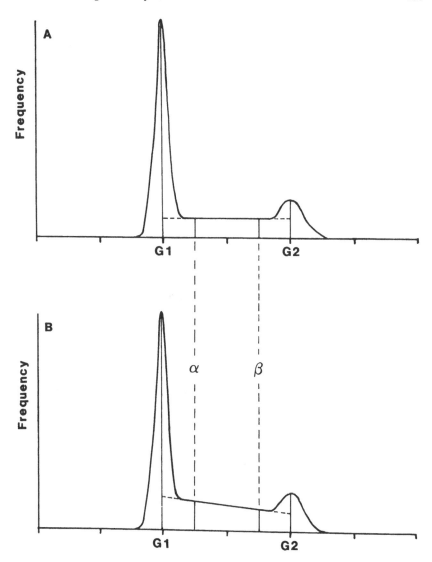

Figure 8.5. Rectilinear integration. Panel A depicts a horizontal top to the S-phase component and panel B depicts an inclined top.

et al. (1975) and is quite adequate for data sets that approximate the shapes of the histograms shown in Figure 8.5. However, the procedure tends to fall apart when the coefficient of variation of the G1 peak is greater than about 5% and where the upper border of the S-phase interval is not flat.

8.3.3 Multiple Gaussian

This method of analysis has been used by Fried (1976, 1977) and by Fried and Mandel (1979). The S-phase interval is modelled with a series of closely

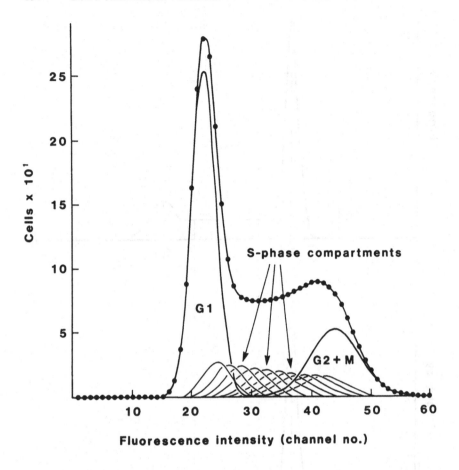

Figure 8.6. Multiple Gaussian analysis (redrawn from Fried 1976).

spaced Gaussian distributions with constant coefficient of variation which are appropriately summed. The principles of the analysis can be appreciated by considering the idealised DNA histogram of Figure 8.2. If the numbers of cells in all channels of this histogram are "spread" according to Gaussian distributions then we obtain the result shown in Figure 8.6. The coefficient of variation is assumed to be constant, hence the standard deviation of the distributions increases with channel number. We can see in Figure 8.6 that the G1 and G2+M peaks are fitted by single distributions, and that the S-phase interval is composed of a number of overlapping distributions which are summed to give the total S-phase distribution. The mean and standard deviation of the G1 and G2+M peaks are estimated from the experimental data, and these are used as starting points to generate the theoretical distribution which is then compared with the experimental data. The multiple parameters are varied until a satisfactory fit is obtained.

8.3.4   Polynomial

A number of different shapes can be generated with polynomial curves described by an equation of the form

$$y = a_0 + a_1 \times x^1 + a_2 \times x^2 + \cdots + a_k \times x^k,$$

where k is the order of the polynomial; see Section 5.2.3. Dean and Jett (1974) introduced this method, and Figure 8.7 is reproduced from their original paper. Panel B shows the experimental DNA histogram as the points and panel A, which is directly comparable with Figure 8.2, shows the "ideal" histogram in which there is no dispersion in the data. In the polynomial analysis the G1 and G2+M peaks are fitted with Gaussian distributions and the upper border of the S-phase interval, between the G1 and G2+M channels in panel A, is fitted by a polynomial. The technique involves estimating the G1 and G2+M modal channels with their standard deviations plus approximate numbers of cells in the G1 and G2+M peaks. This, so far, adds up to a total of six parameters, all of which can be estimated (to a first approximation) from the experimental data. A second-degree polynomial now requires another three coefficients, making a total of nine parameters that are varied until the model fits the experimental data. The curves in panel B were generated by the model, and the predicted S-phase distribution – computed from the polynomial after spreading with constant coefficient of variation – is represented by the dashed curve.

8.3.5   Single Gaussian

A very simple system of analysis has been developed in our department which relies only on the assumption that the data are normally distributed; this is a simplification of the multiple Gaussian technique. The G1/S and the S/G2+M interfaces are modelled by a probability function derived from fitting Gaussian curves to the G1 and G2+M peaks within the region where there is minimal overlap with the S-phase segment of the histogram (Ormerod, Payne and Watson 1987; Watson, Chambers and Smith 1987).

In the perfect data set depicted in Figure 8.2, all G1 cells would be scored in a single channel. Thus, the probability of finding an S-phase cell in the single G1 channel would be zero and the probability of finding an S-phase cell in channel G1+1 would be unity. Therefore, the probability of finding an S-phase cell at the G1/S boundary would be 0.5. In real data sets that are distributed, the probability of finding an S-phase cell at the G1/S boundary is also 0.5, and the boundary must be distributed about the G1 mean channel. The probability function to describe this distribution of the G1/S boundary about the G1 mean channel is modelled as the cumulative frequency curve of the Gaussian which is fitted to the G1 peak. This is illustrated in Figure 8.8, which for simplicity shows only the G1/S interface. The thick curve bounds the whole data set and the dashed curve to the right of the G1 mean represents the G1 component after

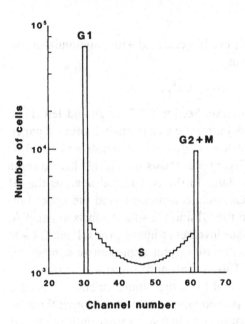

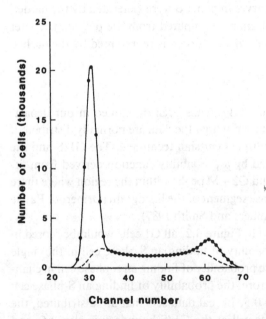

Figure 8.7. Polynomial analysis (redrawn from Dean and Jett 1974).

subtraction of the S distribution. The dot-dashed curve is the cumulative frequency of the G1 distribution, which is arbitrarily scaled to the unit peak height
of the G1 component. This cumulative frequency curve is calculated from the
standard deviation of the G1 peak and represents the probability boundary between the G1 and S components, with a value of 0.5 occurring at the G1 mean.

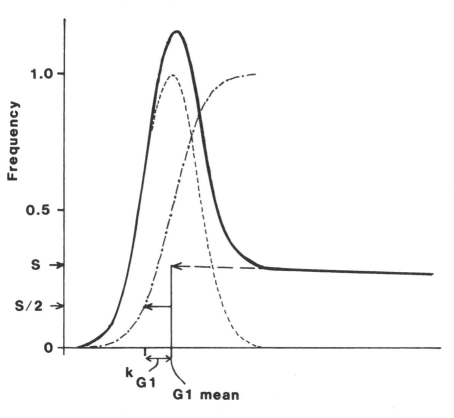

Figure 8.8. G1/S region of a DNA histogram, where the maximum frequency of the G1 component is scaled to unity. The thick uninterrupted curve bounds the whole data set. The dotted curve, mainly to the right of the G1 mean, is the G1 component (Gaussian distributed) and the dot-dashed curve is the cumulative frequency of G1, also scaled to unity. The dashed line with negative slope extrapolates the S-phase envelope above G1 mean + 3SD to the G1 mean channel and cuts the latter at frequency S. The constant k_{G1} is the number of G1 standard deviation units measured from the G1 mean channel associated with a cumulative frequency of S/2.

At $3\sigma_{G1}$ above the mean, where σ_{G1} is the standard deviation of the G1 peak, the cumulative G1 frequency is unity and the respective probabilities of finding a G1 and S-phase cell are zero and unity. Thus, if we know the frequency of the S distribution at the G1 mean, we can calculate the probability within the whole distribution of finding an S-phase cell at the G1 mean from the cumulative frequency distribution of the G1 compartment. An approximation for this S frequency can be obtained by extrapolating the upper border of the S-phase distribution above G1 mean + $(3\sigma_{G1})$ to the value shown as S at the G1 mean using regression analysis. This is shown in Figure 8.8 as the dashed straight line with negative slope. We can now find the point on the cumulative frequency curve associated with a value of S/2, which is the point at which the probability of finding an S-phase cell is 50%. The number of standard deviation units, k_{G1}, associated with this point can now be calculated, which adjusts the position of the G1/S interface within the histogram so that the chance of finding an S-phase cell at the G1 mean will be 50%.

The probability distribution for S phase, P_S, is such that when the frequency in channel x of the data set is multiplied by its corresponding value of $P_S(x)$ we obtain the number of S-phase cells in that channel. The form of this distribution is given by

$$P_S(x) = \int_{-\infty}^{x} \text{erf}(G1(x) - k_{G1}) - \int_{-\infty}^{x} \text{erf}(G2(x) + k_{G2}), \qquad (8.1)$$

where $\text{erf}(z)$ is the error function. $G1(x)$ and $G2(x)$ in equation (8.1) are given by

$$G1(x) = (\text{channel}(x) - G1 \text{ mean})/\sigma_{G1},$$

$$G2(x) = (\text{channel}(x) - G2 \text{ mean})/\sigma_{G2},$$

where σ_{G1} and σ_{G2} are the standard deviations of the G1 and G2+M distributions, respectively. The constant k_{G2} in the S-phase probability function $P_S(x)$ serves the same role at the S/G2+M interface as k_{G1} at the G1/S interface. A pictorial representation of the probability function is shown in Figure 8.9.

In the report by Baisch et al. (1982), a number of synthetic data sets with known properties in G1, S and G2+M were analysed blind by all groups involved in the study. A selection of eight of these data sets, where there was a well-defined G1 peak but with widely differing proportions in the cell-cycle phases, were re-analysed using the method described in this section and are shown in Figure 8.10. Figure 8.11 shows the model predicted proportions in each phase from the analyses in Figure 8.10 plotted against the known proportions.

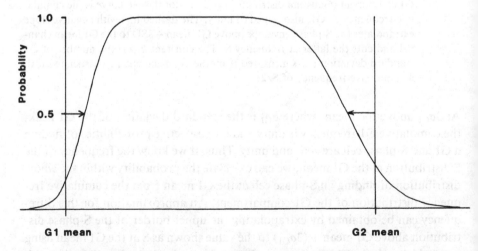

Figure 8.9. S-phase probability distribution. When each channel of this distribution is multiplied by the frequency in the corresponding channel of the experimental data set, the S-phase distribution is generated.

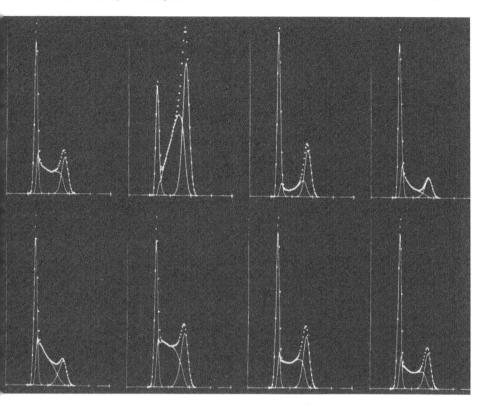

Figure 8.10. Illustration of histogram analysis (simulated data). The dots bound the whole of each data set and the curves bound G1, S and G2+M. Each abscissa division represents 100 channels, and the data sets were scaled individually to the maximum height of each histogram. For display purposes these data have been reduced from 10-bit to 8-bit resolution.

The closed and open circles represent G1 and S respectively and the squares represent G2+M. The line has been drawn with unit slope. $\Sigma \chi^2$ for the 24 comparisons was 5.19 with 15 degrees of freedom, $p < 0.0001$ that the deviations of the predicted from the true could have arisen by chance. Although 24 points were included in the $\Sigma \chi^2$ calculation there are only 15 degrees of freedom, as only two values from each histogram can be assigned arbitrarily; the third value is fixed by the other two. The average of the predicted G1 mean values was 138.2 compared with a true value of 138, an error of less than 0.2%. The G2+M:G1 ratio was 1.993, an error of less than 0.36%. The mean coefficients of variation of the G1 and G2+M peaks were 5.68% and 6.18% respectively, compared with a true value of 5.75% for both.

The technique has now been used for the analysis of over 25,000 histograms within our unit and, with a minor modification, the same technique has been used by Ormerod on a routine basis at the Institute of Cancer Research (Orme-

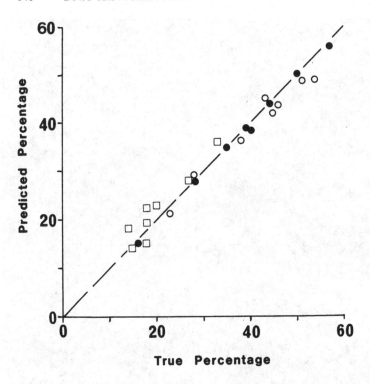

Figure 8.11. Model-predicted percentages plotted against true percentages for the data analysed in Figure 8.10. Solid symbols (•), open symbols (o) and squares (□) represent G1, S and G2+M respectively.

rod et al. 1987). In practice the strength of the method has been found to be the ability of the algorithms to cope with therapeutically perturbed data. It was originally tested with five different human cell lines, HT29 (colon adenocarcinoma), MRC5 (fibroblast line; normal donor), AT5BI (fibroblast line from a patient with ataxia teliangectasia) and SV40 transformed variants (MRC5CVI and AT5BIVA) of the latter cell types. Suspensions were prepared at a concentration of 5×10^6 cells ml^{-1} in full growth medium, and the cells were stained with triton X-100/ethidium bromide (Taylor 1980). In this initial evaluation of over 350 DNA histograms from the five different cell lines, in both control and drug- or radiation-perturbed populations the G2+M:G1 ratio was 2.009 with 95% confidence limits of 0.015. This overall mean value does not differ significantly from the expected ratio of 2.0, $p > 0.05$. There was somewhat greater variability in the subgroup of drug- and radiation-perturbed data, where the mean was 2.014 with 95% confidence limits of 0.021. The computed proportions in G1, S and G2+M varied within the ranges of 10% to 90%, 7% to 90% and 5% to 80% (respectively) in a multiplicity of combinations. A selection of six histogram analyses is shown in Figure 8.12 to illustrate the wide variety of forms that can be analysed by the model.

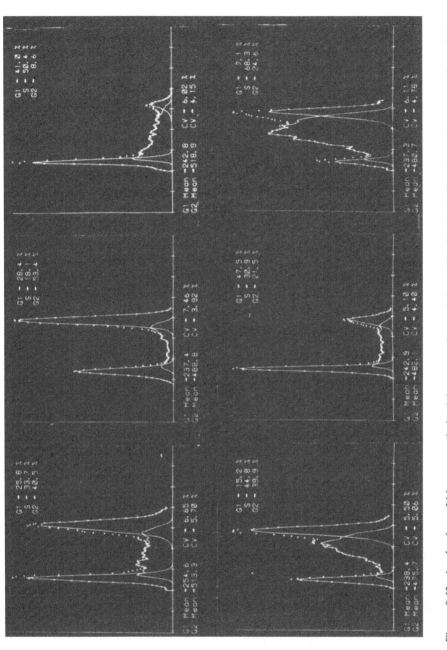

Figure 8.12. A selection of histograms derived from experimental data. The display is directly analogous to Figure 8.10.

These analyses look reasonable; however, there is no direct evidence that the proportions of cells in the three phases from the experimentally derived data are correct. Indeed, direct evidence is almost impossible to obtain because there is no completely independent method of estimating the proportions in G1 and G2 directly, and in perturbed populations the S-phase fraction cannot be reliably estimated. In completely unperturbed populations the ^{3}H-thymidine labelling index gives a good approximation for the proportions in S. However, in perturbed data sets (the majority in our initial evaluation) the labelling index must never be equated with the proportions in S, as a cell may have an S-phase DNA content but not be synthesizing DNA owing to the perturbing event. The majority of the experimental data sets exhibited a well-defined G1 peak and the average G2+M:G1 ratio for over 350 analyses was 2.009, very close to the expected value. This is comparable to the result from the simulated data, and the range of shapes of the experimental data was very similar to that found in the simulated histograms. We have no reason to suspect that any errors in the analysis of the experimental data were either qualitatively or quantitatively different from those in the simulated histograms; we thus conclude that analysis of the former gives, at least, reasonable results.

In order to test the algorithms further, a series of experiments was carried out with duplicate flasks. Figure 8.13 shows the proportions in G1, S and G2+M from the first sample plotted against the comparable values from the second sample. The closed and open circles depict G1 and S, respectively, and the squares represent G2+M. A total of 75 points is shown in Figure 8.13 but the regression line was calculated from a total of 168 points, which gave a slope of 0.97 and intercept of 1.04 with 95% confidence limits of 5.2%. The correlation coefficient was 0.962. A $\sum \chi^2$ calculation for the deviation of one measurement from the other yielded 46 with 111 degrees of freedom, $p < 0.0001$ that the differences could have arisen by chance. The result in Figure 8.13 indicates that the true proportion within a phase will have a 95% chance of being within ±5.2% of the computed value, and most investigators would accept a 5% biological variability between experiments. The duplicate samples analysed here contain more potential sources of variation than the originals owing to staining, instrumental factors and the analysis procedure. These samples were duplicated from the very start of each experiment and so additional sources of variation must be considered. These include dilution errors during seeding, growth variation from flask to flask, variations in radiation dose or drug concentration and exposure time (about 70% of these duplicates were perturbed populations), versene artefacts in single-cell suspension preparation, dilution errors in preparing the suspension for staining, and possible modulation of the staining by the perturbing treatment under study. When all of these factors are taken into consideration it would seem unreasonable to expect reproducibility to be better than ±5% between samples, and re-runs of

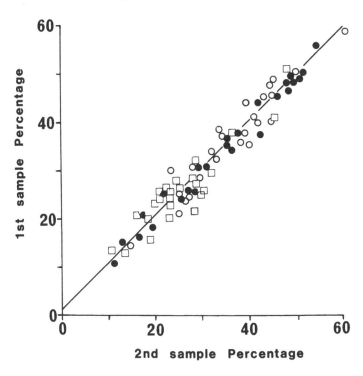

Figure 8.13. Analysis of duplicate samples with symbolic representation as in Figure 8.11. Data from only 25 analyses (75 points) are shown although the regression was calculated from 56 data sets (168 points). The slope and intercept were 0.97 and 1.04 respectively, with a correlation coefficient of 0.962.

the same samples gave results for the proportions within each phase to within ±1.5%.

We have found this method of analysis to be superior to that based on age distribution theory published previously (Watson 1977b). The model is considerably more robust, and it can cope quite adequately with populations where there is a varying rate of increase in DNA during S phase, as well as with perturbed data sets.

8.3.6 TCW analysis

TCW stands for trace, cut and weigh, and can be tried in emergencies when all else fails. It is not seriously recommended and it works only with fairly standard-looking data sets. It depends on the human mind's exceptionally good analogue processing, pattern recognition and discriminatory capacity. The histogram is covered by high-quality tracing paper, and the sections of the histogram estimated to correspond to the G1, S and G2+M areas as in Figure 8.12 are cut out with a scalpel and weighed on a chemical balance. The weights are then proportional to the proportions in the three cell-cycle phases. I have used this method in the past and took the trouble to calibrate the tracing paper, where different known areas were weighed to produce a linear calibration. The results for the phase proportions were within ±3% of those obtained by computer model where I was able to compare the two.

9

Cell-cycle kinetics

The type of histogram data and its analysis discussed in Chapter 8 can yield the proportions of cells in the various cell-cycle phases and the relative durations of the phases only if there is exponential growth and all the cells are within the division cycle. This is not kinetic information, even though some publications might lead one to believe that it is. Kinetic information can be obtained only when a perturbing event or second marker is introduced into the steady state and where the effects thereof are followed temporally. Analysis of the decay of the introduced event with time yields the kinetic information. There are two basic types of such perturbed DNA histogram. The first is due to an alteration in the steady-state age distribution with partial synchrony of the population. The second is more subtle, where a specific marker for S phase is introduced without altering the steady-state age distribution of the population.

9.1   Stathmokinetic techniques

An alteration of the steady state of growth may be induced by addition of drugs or radiation which block or delay the population in a specific phase of the cell cycle. Addition of colcemid, which depolymerises tubulin in microtubules, induces metaphase arrest as cells cannot proceed to anaphase. Figure 9.1 shows DNA histograms of a human breast cancer–cell line before (panel A) and after (panel B) treatment with colcemid, where the accumulation of cells in G2+M is obvious. Rasoxane (ICRF 159) is another drug that arrests cells with a G2+M DNA content, and the data shown in Figure 9.2 were obtained from a mouse tumour cell line blocked at time zero. The population was sampled at intervals and a progressive shift to the right is noted. A sequential series of data sets such as these can be analysed to give the absolute cell-cycle time, the phase durations and their standard deviations.

One major objection to this stathmokinetic approach is that the perturbing event may alter the behaviour of the population in ways other than those that are expected and are being observed. It was during such a series of experiments

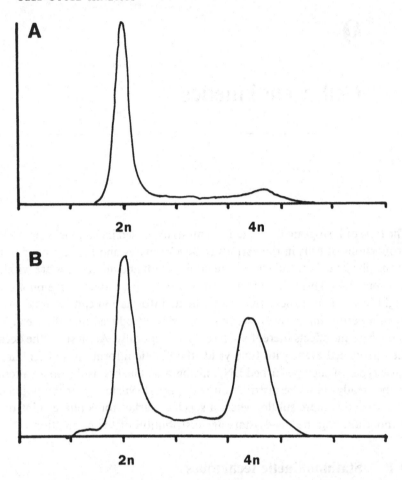

Figure 9.1. DNA histograms before (A) and after (B) treatment with colcemid, showing accumulation of cells with G2+M DNA content.

with stimulated B cells using colcemid to block cells in G2 that we discovered that this also causes a block in early G1 (Kenter et al. 1986; see also Section 9.8).

9.2   Mitotic selection

A second approach, which goes some way towards overcoming these objections, involves mitotic selection (Teresima and Tolmach 1963). Cells growing in monolayer cultures are usually fairly firmly attached to the substratum of the containing vessel and are spread over a relatively large area. However, at mitosis the cells become spherical and a relatively small area of the cell wall is in contact with the plastic or glass, so the mitotic cells can be selectively detached by gently shaking the flask. The culture medium, containing detached mitotic

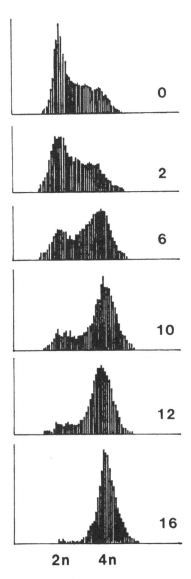

0

2

6

10

12

16

2n   4n

Figure 9.2. Rasoxane (ICRF 159)–treated mouse tumour cells showing a progressive increase in cells arrested with a G2+M DNA content.

cells, is poured off, the cells are concentrated by centrifugation and re-seeded into a number of fresh flasks. Within a few minutes the cells have completed division, and they stick down to the surface as G1 cells immediately following mitosis. At intervals thereafter a flask of cells is stained for DNA content.

The results of one such experiment are shown in Figure 9.3. At 2 hours after selection there was a clear G1 peak. By 4 hours a small proportion of cells

POST SYNCHRONISATION PROGRESSION of EMT6/M/CC CELLS

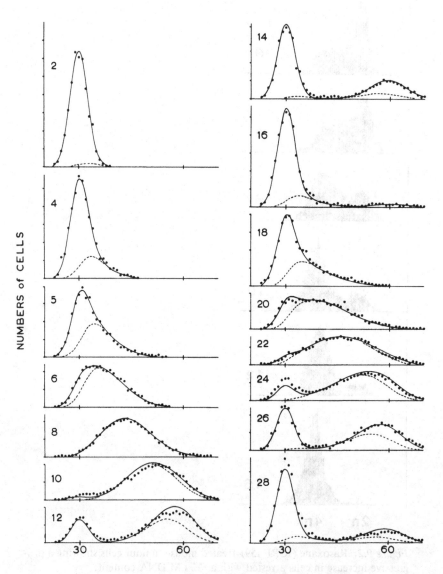

DNA FLUORESCENCE INTENSITY

Figure 9.3. Mitotic selection synchrony experiment showing a clear progression of cells through two cell cycles after re-seeding. The continuous curves bounding the whole of each distribution and the dashed curves bounding the S-phase components were predicted by computer model.

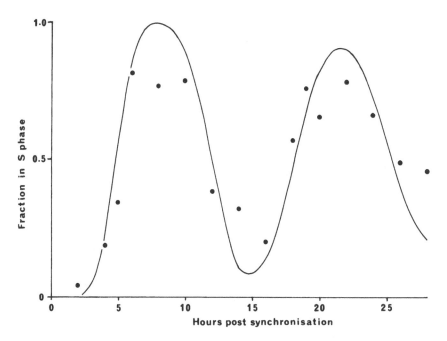

Figure 9.4. Tritiated thymidine labelling index (points) compared with proportion in S phase predicted by computer model analysis (curve) of the data
shown in Figure 9.3.

has increased their DNA content, as they began synthesising DNA in S phase.
Thereafter there was a progressive shift to the right, and with time a "wave"
traversed the S-phase segment of the histogram. At 12 hours some cells had
progressed through the cycle and had divided to re-appear with a G1 DNA
content. There was then a second wave, as cells again moved into synthesis to
be followed by a second division. These data were analysed by computer model
(Watson and Taylor 1978) as described in Section 9.3 to give a cell-cycle time of
13.3 hours. The intermitotic phase times were 4.8, 7.2 and 1.3 hours respectively
for G1, S and G2+M. The tritiated thymidine labelling index was also determined at each time point, and these data were compared with the computed
fraction in S phase shown in Figure 9.4. There was generally good agreement
between the computed curve and experimentally determined labelling index,
apart from a discrepancy at the first peak where the labelling index is probably
underestimated.

9.3   Modelling population kinetics

Two classes of problems arise when modelling DNA histogram population kinetics. The first pertains to the dynamic processes that must be described

in the form of time-dependent mathematical functions. These functions are designed to elucidate how the proportions of cells within the three subphases (G1, S and G2+M) definable by flow cytometry vary with time. The second class pertains to how the proportions in the cell-cycle phases so calculated are expressed as a DNA histogram. Some of the processes involved under the second class have already been covered in Chapter 8, where the proportions in G1 and G2+M are modelled with Gaussian distributions. However, a type of system different from that described in Chapter 8 must be used to model the S-phase component, as will be discussed after first considering the dynamics.

Many of the mathematical models used in flow cytometry to describe population dynamics take their origin from work carried out in the mid to late 1960s to model PLM (percent labelled mitosis) curves (Hartmann and Pederson 1970; Takahashi 1966, 1968; Takahashi, Hogg and Mendelsohn 1971). Those that can readily be adapted for use in flow cytometry fall into two basic categories. The Takahashi–Hogg–Mendelsohn model divided the cell cycle up into a number of arbitrary compartments, each of which represented a time domain. For example, if the cell-cycle time were 10 hours and if this were divided into 20 equal compartments then each would be of 0.5 hours' duration. If we assume that the durations of G1, S and G2+M were 4, 5 and 1 hours respectively then 8, 10 and 2 arbitrary compartments would be assigned to G1, S and G2+M. This is depicted in the top panel of Figure 9.5, an exponential age distribution where the S-phase cells are cross-hatched and the G2+M cells are stippled. At a time t later, where $t = t_{G1}/2$, the whole population will have moved four compartments (half of G1), or time domains, to the right. This is shown in the bottom panel of Figure 9.5, where the cells originally in G2+M have divided and progressed through the cell cycle to occupy time domains 3 and 4 in G1. Cells originally in the last two time domains of S phase (i.e. late S cells) have progressed through G2+M and now occupy the first two compartments of G1. The remainder of the original S-phase cells now reside in the time domains of mid to late S and G2+M, and half the G1 cells have progressed into S phase.

The data obtained for PLM curves consist of the fractions of mitotic cells autoradiographically labelled with silver grains at known times after flash labelling with ^{3}H-thymidine. In the top panel of Figure 9.5 the PLM would be zero; however, in the bottom panel 100% of mitoses would be labelled, as the whole of the G2+M compartment is now occupied by cells that were in S phase when the ^{3}H-thymidine was introduced into the system. A typical PLM curve is shown in Figure 9.6, where the approximate durations of G1, S and G2+M are indicated and the damping is due to variability in the phase time durations. In order to analyse these types of data we must assume that some type of mathematical function(s) can simulate these processes, and that these must be able to describe not only the phase durations but also the dispersion in those durations. Various investigators have used log-normal distributions, gamma functions, and Poisson and normal distributions to obtain an approximation to the

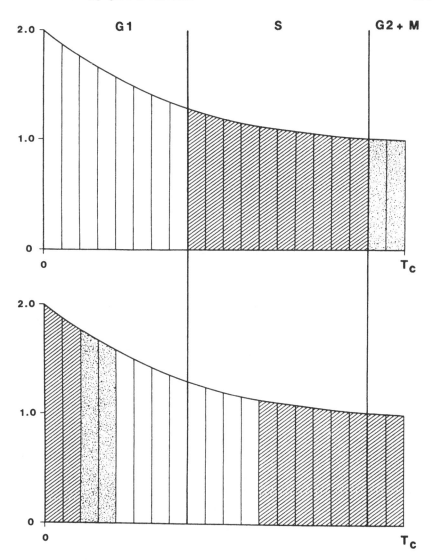

Figure 9.5. Exponential age distribution with 20 equal time compartments, each of 0.5 hours duration (top panel). The bottom panel depicts the progression of cells through four time domains to the right, half of the duration of t_{G1}.

mean phase durations and their variances. In general it doesn't seem to matter very much which distribution is used in PLM analysis, as they all give very similar results for the phase times.

Joe Gray (1976) at the Lawrence Livermore Laboratories has adapted the Takahashi–Hogg–Mendelsohn model for flow cytometric analysis of sequential DNA histograms assuming Gaussian distributions for the phase times, which gives the mean durations with their standard deviation. The transit of cells

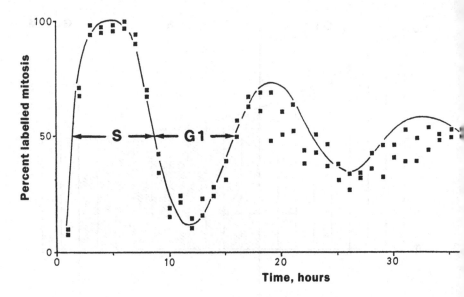

Figure 9.6. Typical PLM (percent labelled mitosis) curve, which is damped owing to variation in the phase times. The points represent the experimental data and the curve was fitted by computer model. The durations of G1 and S are indicated, and the duration of G2+M is approximately from time zero to the point at which the first wave reaches its half height.

from one compartment to the next is described by a series of sequential differtial equations, one for each compartment interface. This is a very robust model, and perturbed populations can be readily analysed by simulating the experimental conditions within the model. For example, if the population had been treated with an agent that kills all cells in S phase (e.g. hydroxy-urea) then the starting conditions for the model would be adjusted to include the absence of cells in the S-phase domains of the age distribution diagram shown in the top panel of Figure 9.5. After mitotic selection all cells would be assumed to occupy the first time domain of G1 but, where cells had been treated with a G2+M arresting agent (colcemid and rasoxane), the age distribution would commence as shown in the top panel of Figure 9.5 but in the mathematics no cells would be "allowed" to progress beyond the last time domain of G2+M. Gray's model also contains the added sophistication that the rate of traverse through S phase need not be constant, hence variable DNA synthesis rates can be accommodated. However, this method does have a practical disadvantage, stemming from a restraint in the original concept (Takahashi 1966, 1968), in that the phase-time coefficients of variation (CV) can be varied only in discrete steps. The step size is inversely proportional to the number of arbitrary subcompartments that a particular phase is divided into. Thus, in Figure 9.5 the minimum effective standard deviation for G2+M would be 0.5 hours and, since this phase has a duration of 1.0 hours in the diagram, the minimum CV would be 50%.

In order to obtain smaller CVs the number of subdivisions must be increased, which increases the size of the differential equation matrix and consequently increases the computing time. Although Gray used the fast "predictor–corrector" method of Hamming (1973) to solve the differential equation matrix, the computing power needed for small CVs is considerable, particularly for data sets containing a large number of DNA histograms.

The second approach to modelling PLM curves which can be readily adapted for flow cytometry is due to Hartmann and Pederson (1970), where a single analytic equation is obtained for each of the three cell-cycle phases (Watson and Taylor 1978). Consider the normal distributions in Figure 9.7, where t_{G1} is the mean duration of G1 with a standard deviation of σ_{tG1}. Let us assume that

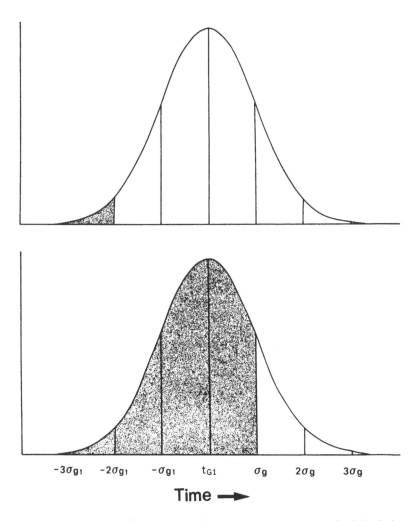

Figure 9.7. Normal distribution with mean of t_{G1} and standard deviation of σ_{tG1}.

cells initially were synchronised at the beginning of G1, which is time zero on the abscissa of the diagram. If we place a piece of card to cover the diagram with the left-hand edge at time zero and progressively move this to the right, we will stimulate the passage of cells through G1. All cells will remain in G1 until the left-hand edge of the card hits the curve's left shoulder, which begins at $t_{G1} - 3\sigma_{tG1}$. As the card is moved further to the right (increasing time), a greater proportion of the area under the curve will be exposed. When the edge of the card hits the point $t = t_{G1} - 2\sigma_{tG1}$, approximately 3% of the total area under the curve will have been exposed (top panel). As we move still further to the right, half the area will be exposed at $t = t_{G1}$ and 82% will be exposed at $t = t_{G1} + \sigma_{tG1}$ (bottom panel). The shaded areas in both panels represent the proportions of the whole population which have completed G1 at times $t_{G1} - 2\sigma_{tG1}$ and $t_{G1} + \sigma_{tG1}$ respectively. The total area under the normal curve is normalised to unity; thus, at $t = t_{G1} + 3\sigma_{tG1}$, 100% of the cells will have completed G1.

The error function, erf(Z), is used to compute the area under the normal curve from $-\infty$ to Z; in this context $Z = (t - t_{G1})/\sigma_{tG1}$. Thus, the probability of a cell having exited G1, EXITG1, at time t is given by

$$\text{EXITG1} = \text{erf}((t - t_{G1})/\sigma_{tG1}).$$

As cells progress from G1 to S the exit probability from G1 must be the entry probability to S, so the probability of a cell still being in G1 is given by $1 - \text{EXITG1}$. Thus, we can express the transit through the cell cycle as a series of three probability functions PG1, PS and PG2 which respectively give the probability of cells being in G1, S and G2+M:

$$\text{PG1} = 2 \times \text{EXITG2} - \text{EXITG1},$$

$$\text{PS} = \text{EXITG1} - \text{EXITS},$$

$$\text{PG2} = \text{EXITS} - \text{EXITG2}.$$

The entry probability for G1 must be twice the exit probability from G2+M in order to account for cell division. The full equation complex describing these functions mathematically may be given as

$$\text{PG1} = \sum_{I=K}^{I=N} \left\{ \text{erf}\left[\frac{2[t - (t_c \times J)]}{\sqrt{(\sigma_{tc}^2 \times J)}}\right] - \text{erf}\left[\frac{t - t_{G1} - (t_c \times J)}{\sqrt{[\sigma_{tG1}^2 + (\sigma_{tc}^2 \times J)]}}\right] \right\},$$

$$\text{PS} = \sum_{I=K}^{I=N} \left\{ \text{erf}\left[\frac{t - t_{G1} - (t_c \times J)}{\sqrt{[\sigma_{tG1}^2 + (\sigma_{tc}^2 \times J)]}}\right] - \text{erf}\left[\frac{t - t_{G1} - t_S - (t_c \times J)}{\sqrt{[\sigma_{tG1}^2 + \sigma_{tS}^2 + (\sigma_{tc}^2 \times J)]}}\right] \right\},$$

$$\text{PG2} = \sum_{I=K}^{I=N} \left\{ \text{erf}\left[\frac{t - t_{G1} - t_S - (t_c \times J)}{\sqrt{[\sigma_{tG1}^2 + \sigma_{tS}^2 + (\sigma_{tc}^2 \times J)]}}\right] - \text{erf}\left[\frac{t - (t_c \times I)}{\sqrt{(\sigma_{tc}^2 \times I)}}\right] \right\},$$

where $J = I - 1$.

The equation complex just listed describes the kinetics within the population, and we now have to translate the proportions in the cell-cycle phases into a DNA histogram. The proportions in G1 and G2+M are modelled as normal distributions and the S-phase fraction is generated as follows. If we assume that the rate of DNA synthesis is constant (or nearly constant) then the mean rate at which cells traverse the S-phase interval of the DNA histogram is given by $(M-1)/t_S$ channels per unit time, where M is the mean channel of the G1 peak. The "effective mean channel" of the S-phase distribution (EMS) is given by

$$\text{EMS}(I) = M + \left\{ \frac{M-1}{t_S} \{ t - (t_{G1} + (t_c \times J)) \} \right\},$$

where $J = I - 1$ and I varies from K through N cycles. It can be seen that $\text{EMS}(1) = M$ for $t = t_{G1}$, at which point half the G1 cells will have entered S phase. This gives the position of the distribution; the shape is computed with a channel variance of $S(I)^2$, where

$$S(I)^2 = \left\{ \frac{M-1}{t_S} \right\}^2 [\sigma_{tG1}^2 + \sigma_{tS}^2 + (\sigma_{tc}^2 \times J)].$$

This accounts for the spreads in the G1 and S intermitotic phase times. Only that portion of the S distribution lying within the interval $M+1$ to $2M-1$ channels inclusive represents cells in S, irrespective of the values of $\text{EMS}(I)$ and $S(I)$. The distribution so obtained is now "spread" on a channel-by-channel basis to account for the increasing flourescence standard deviation attributable to the constant CV, and the final distribution is generated with the appropriate summation.

9.4   Multiparameter optimisation

The kinetic cell-cycle equation complex shown in Section 9.3 has six unknown parameters: t_c, t_{G1}, t_S, σ_{tG1}, σ_{tS} and σ_{tG2}. The duration of G2+M, t_{G2}, is not unknown as this is fixed by t_c, t_{G1} and t_S, and likewise the standard deviation of the whole cell cycle, σ_{tc}, is fixed by σ_{tG1}, σ_{tS} and σ_{tG2}. Two further parameters, the G1 mean flourescence channel and its CV, can be obtained from the experimental data.

The data shown in Figure 9.3 were simulated with the model as follows. Initial estimates for the cycle time and the relative phase durations were obtained by inspection of the data and by analysis of an exponentially growing control population. The cell-cycle time was obviously between 12.5 and 14.5 hours, and the relative phase durations were estimated to be between $0.3t_c$ and $0.4t_c$, $0.5t_c$ and $0.6t_c$ and between $0.05t_c$ and $0.15t_c$ for G1, S and G2+M respectively. The 2-hour point was analysed to give a G1 mean fluorescence channel of 30 with standard deviation of 2.5 channels. The coefficients of variation of the phase durations were guessed as being between 10% and 30%. The parameters were varied iteratively, and for every combination the computer-predicted distributions were compared with the experimental data. The curves in Figure 9.3 were

Table 37. *Combination of parameter values giving the best fit between the mathematical model and the experimental data*

Phase	Time	CV
t_{G1}	4.8	0.23
t_S	7.2	0.16
t_{G2}	1.3	0.28
t_c	13.3	0.12

generated with the combination of parameters that gave the best fit, using the simple sum of deviations as the criterion, where the dashed curves within each histogram represent the computed S-phase distribution. The results for the phase durations with their coefficients of variation, CV, are shown in Table 37.

Purely iterative computing techniques are robust but not very efficient. In the preceding example, the parameter value ranges were initially divided into 5 incremental levels; since there were 6 parameters, this required a total of 5^6 (15,625) iterations. In the second phase the ranges were divided into 4 incremental levels, which required a further 4^6 (4,096) iterations. After "fine tuning" for some of the less well-conditioned parameters, just over 20,000 iterations were needed to obtain the results shown in Table 37.

Parameter optimisation is all about getting the "answers" (to within specified limits) in as few iterations as possible. Finding the best fit between a multivariable model and experimental data is most easily understood if the arguments are developed from a single-variable problem.

9.4.1   Conventional space

Let us suppose that y is a function of the single variable a. The mathematical notation is

$$y = f(a).$$

Let us further suppose that an analytical solution cannot be found. A practical example of one such equation is

$$\mathrm{LI} = \exp\left(-\left(\frac{\log_e(2)}{T}(t_{G1} + t_S)\right)\right) - \exp\left(-\left(\frac{\log_e(2)}{T}t_{G1}\right)\right), \qquad (9.1)$$

where LI is the labelling index, T is the potential doubling time and where t_{G1} and t_S are the durations of G1 and S phase respectively (Steel 1968). This is an important equation in human tumour kinetic work (see Section 9.7); that's why I've chosen it for this example. Although the potential doubling time (T) is the only unknown in this equation, we cannot produce an analytic solution because we cannot obtain T in terms of the remaining parameters. However, if we denote the whole of the right-hand side (RHS) of equation (9.1) by y' and rearrange, we obtain

$$LI - y' = 0.0. \tag{9.2}$$

We can now guess a value for T, evaluate the RHS of equation (9.1), subtract this from the experimentally determined value of LI and compare the result with zero. If the initial guess for T is incorrect then the valuation of equation (9.2) will not be zero, and this non-zero value is denoted the *error*. The units of the error in a 1-dimensional problem such as this are of no consequence, as we are concerned only with its magnitude and sign. A second guess for T is made, and this too will have an attendant error. If the error associated with the second guess is greater than that associated with the first, we have moved T in the wrong direction. Thus, we now know in which direction to alter T, and the third trial value is obtained by moving T by twice the initial increment in the opposite direction to that of the initial move. We continue moving T in this direction while the error is decreasing and there is no change in sign. As soon as the sign of the error changes we have overshot the optimum value, so the increment is halved and the direction of movement is reversed. Thus, by a series of alternations of the sign of the error with decreasing increments of T, we successively minimise the deviations of the theoretical result from the observed. This process is illustrated in Figure 9.8, which shows the error plotted against the sequence of approximations of T. It is clear from our discussion that in order to optimise this single-variable function it is necessary to set up and evaluate two initial values of T, so that the direction of subsequent movement can be determined. Furthermore, it is the value with the greater error which is moved, and after any given move there are always two values with their associated errors for comparison.

Let us now extend these principles to the two-variable function

$$y = f(a, b),$$

where y is dependent on both a and b. Any given pair of values (a_n, b_n) which is the coordinate of a point in 2-dimensional space will be associated with an error after evaluation of the function. We have already seen that to determine the direction of movement in a 1-dimensional problem it is necessary to calculate using two initial values. Similarly, with a 2-dimensional problem we must evaluate three initial pairs of values, (a_1, b_2), (a_2, b_1), (a_3, b_3), which form the three points at the corners of a triangle. These points will be assigned respective identifying numbers 12, 21 and 33 with errors of 20%, 25% and 10% respectively. Point 21 is associated with the greatest error and must be moved to a lower error region. On the information available this would seem to be somewhere beyond a line joining points 12 and 33. We can now calculate a point on

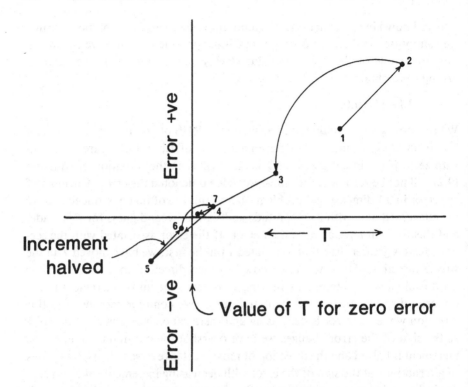

Figure 9.8. "Error" from a 1-dimensional function plotted against the variable *T*. The initial guess is labelled 1, and subsequent moves are shown by the numbers against each point and by the arrow-heads.

the line (12, 33) with coordinates (a_H, b_H) which is weighted in inverse proportion to the errors on points 12 and 33. By taking moments we get

$$a_H = \frac{a_1 e_{33} + a_3 e_{12}}{e_{33} + e_{12}}, \qquad b_H = \frac{b_3 e_{33} + b_2 e_{12}}{e_{33} + e_{12}},$$

where e_{12} and e_{33} are the errors associated with points 12 and 33. The point (a_H, b_H) will be referred to as the "*H*-point" for reasons that will become apparent in Section 9.4.2, and the direction of movement of point 21 will be along the vector defined by joining this point to the *H*-point. The distance moved, *D*, will be double the distance between point 21 and the *H*-point, and by Pythagoras's theorem this is given by

$$D = 2 \times \sqrt{(a_2 - a_H)^2 + (b_1 - b_H)^2}.$$

This process is illustrated in Figure 9.9A, which shows hypothetical error contours within the 2-dimensional (a, b)-space, the initial three points, the *H*-point and the new position of point 21 (which is identified by the square). The error associated with the new position of point 21 (which for ease of reference maintains its initial identifying number) is now less than that of point 12. Thus,

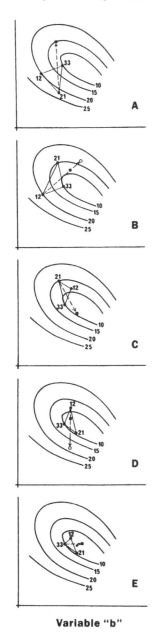

Variable "b"

Figure 9.9. 2-dimensional error contour maps for a two-variable function. The panels A to E show the succession of moves which locate the point of minimum error (see text).

we must now move this latter point, and this is accomplished by the same process as shown in Figure 9.9B. Although this new position of point 12 (identified by the open circle in Figure 9.9B) has a lower error than its original position, it still has an error greater than that associated with either of the other two points, and we must make a second successive move with this point. However, if we merely follow the same procedure again we would move point 12 back to its original position, which gets us nowhere. As the second position of 12 is better than the first we stay in this domain (i.e. above and to the right of the line 21,33) but halve the distance moved from the H-point. This new position is shown as the solid square in Figure 9.9B. We now find that point 21 again has the greatest error, and this is moved to its new position as illustrated in Figure 9.9C. Following this move point 12 is again associated with the greatest error, and it is moved to the position marked by the open circle in Figure 9.9D. This is a worse position, so we make a second successive move with this point. As we moved into a domain of greater error, the distance moved from the H-point is halved and the new position (marked by the solid square in Figure 9.9D) is set up in the domain of smaller error. Point 33, which until now has remained fixed, is now associated with the greatest error and is moved to its new position marked by the solid square in Figure 9.9E.

Thus, after a total of seven moves, the point of minimum error (identified by the asterisk) is within the triangle of points. By continuing these sequences of moves the three points "collapse" inwards towards the point of minimum error. The accuracy of final location is determined by the number of halvings allowed of the distance moved from the H-point, by a limit placed on the difference between the maximum and minimum error associated with the points giving the most extreme values of the error, or by a specified limit on the amount of possible change that can occur for a given parameter. When this specified limit (usually 0.1% of the value of that parameter at the best point) is reached, the optimiser assumes that convergence of that variable has taken place; the value is then fixed and the search process is reduced by one dimension.

The optimisation of the variables in the 3-dimensional function

$$y = f(a, b, c)$$

follows exactly the same principles as described for two dimensions, except that four initial 3-dimensional points are set up. The geometrical shape defined by these points is now a pyramid as opposed to the triangle in the 2-dimensional problem. Again, it is the worst point which is moved, but the move is now through a plane each time and not through a line. The H-point is in the plane opposite the worst point and is calculated from the coordinates of the points defining that plane, taking into account the inverse error weighting of each point. The distance moved by the worst point through the H-point is again calculated by Pythagoras's theorem (extended to three dimensions), where

$$D = 2 \times \sqrt{(a_2 - a_H)^2 + (b_1 - b_H)^2 + (c_w - c_H)^2}. \tag{9.3}$$

The subscripts w and H of the a, b and c coordinates refer to those of the worst point and the H-point respectively. A diagrammatic representation is given in Figure 9.10, but only one move is shown because all subsequent moves are directly analogous to those described for the 2-dimensional problem.

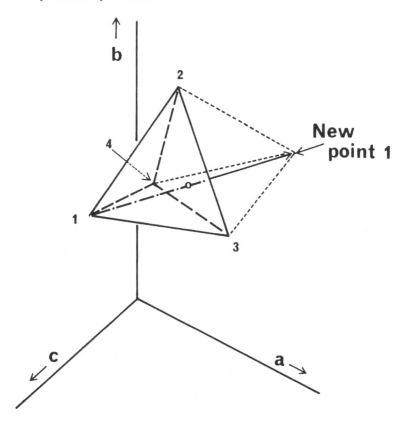

Figure 9.10. Three-variable (a, b, c)-space containing four initial points defining a pyramid marked by points 1 to 4. Point 1 is that with the greatest error and this is moved to "new point 1" through the plane defined by points 2, 3 and 4. The point in the plane through which point 1 is moved is the H-point, shown by the small circle.

9.4.2   Euclidean hyperspace

We are now in a position to generalise the geometry of the optimising process to n-dimensional Euclidean hyperspace, E^n, where $n > 3$ and where the distance between any two points with n coordinates can be defined. The equation of a line in 2-dimensional space is given by

$$y = mx + c.$$

Multiplying through by z we obtain

$$zy = zmx + zc.$$

Let α_1, α_2 and β represent the constants z, zm and zc (respectively), and let y and x be represented by x_1 and x_2 (respectively). On rearrangement we get

$$\alpha_1 x + \alpha_2 x_2 = \beta. \tag{9.4}$$

Equation (9.4) may also be written as

$$[\alpha_1 \alpha_2] \begin{bmatrix} x_1 \\ x_2 \end{bmatrix} = \beta, \tag{9.5}$$

and the line representing this equation can be regarded as the set of all points (x_1, x_2) that satisfy the equation.

Let us consider two points A' and B' in 3-dimensional space, with coordinates (a_1, b_1, c_1) and (a_2, b_2, c_2) respectively. Representing these coordinates as column vectors, we have

$$A' = \begin{bmatrix} a_1 \\ b_1 \\ c_1 \end{bmatrix} \quad \text{and} \quad B' = \begin{bmatrix} a_2 \\ b_2 \\ c_2 \end{bmatrix}.$$

The square of the difference, d^2, between A' and B' is given by

$$(A' - B')(A' - B') = d^2.$$

Hence

$$(a_1 - a_2)^2 + (b_1 - b_2)^2 + (c_1 - c_2)^2 = d^2. \tag{9.6}$$

This equation is the extension of Pythagoras's theorem to 3-dimensional coordinates in E^3 (compare with equation (9.3)). Equations (9.4) and (9.5) can also be extended to three dimensions by adding the term $(\alpha_3 x_3)$ to the left-hand side and generally, therefore, we can extend the notation to n dimensions, where

$$\alpha_1 x_1 + \alpha_2 x_2 + \alpha_3 x_3 + \cdots + \alpha_n x_n = \beta,$$

$$[\alpha_1 \alpha_2 \alpha_3 \cdots \alpha_n] \begin{bmatrix} x_1 \\ x_2 \\ x_3 \\ \vdots \\ x_n \end{bmatrix} = \beta. \tag{9.7}$$

The set H of points that satisfy this equation constitutes a hyperplane in n-dimensional space, and thus equation (9.7) is the equation of that hyperplane. For $n = 3$ the hyperplane is a plane in conventional E^3, if $n = 2$ the hyperplane contracts to a line and to a point if $n = 1$.

Equation (9.6) can also be generalised to n-dimensional space, where the squared distance d^2 between any two points x' and y' is given by

$$d^2 = \sum_{i=1}^{i=n} (x_i' - y_i')^2, \tag{9.8}$$

where $x_i' = (x(1, i), \ldots, x(n, i))$ and $y_i' = (y(1, i), \ldots, y(n, i))$.

Because both the relationship (equation (9.7)) and also the distance (equation (9.8)) between any two points in Euclidean hyperspace have been defined, we can treat the $(n > 3)$-dimensional optimising problem in exactly the same way as described for one to three dimensions. We have already seen that $n + 1$ initial points must be set up for n variables, and we must also include the obvious provision that no two initial points can have identical coordinates. The coordinates for the H-point, so called because it represents a point in a hyperplane in n-dimensional space, can be calculated since the concept of distance in hyperspace has been defined. In common with all other points in the set, the H-point will have n coordinates for n variables. For $n = 4$, where point 5 is that with the maximum error, each weighted coordinate h_i of the H-point can be calculated as

$$h(i=1,4) = \frac{z(1,i)e_2e_3e_4 + z(2,i)e_1e_3e_4 + z(3,i)e_1e_2e_4 + z(4,i)e_1e_2e_3}{e_2e_3e_4 + e_1e_3e_4 + e_1e_2e_4 + e_1e_2e_3}.$$

where the subscripts i each refer to a given coordinate (or parameter) in the Z-plane of 4. The numerical subscripts identify each point, and e is the error associated with these respective points.

In order to illustrate the performance of the optimising routine, a number of synthetic data sets were generated with the parameters in Table 37. A total of five data sets were synthesized, where four of these assumed cell-cycle blocks (stathmokinetic analysis) in early/mid G1, early S phase, mid S phase and mitosis in an exponentially growing asynchronous population. DNA histograms were generated at 3, 6, 9 and 12 hours after the block commenced. The fifth data set contained nine histograms generated at three hourly intervals (3 through 27 hours) after synchronisation of the population in early G1, which approximates to conditions a short time after mitotic selection.

In order to simplify the illustration, only four parameters – the cycle time and CVs of the intermitotic phase times – were computed, as the durations of the intermitotic phase times in relation to the cycle time can be obtained by analysis of an asynchronous exponentially growing population (Gray 1976; Watson 1977b). Hence these parameters were not included; the position of the G1 DNA mean (channel 30) and its standard deviation (2.5 channels) were also excluded, and it was assumed that the G2+M:G1 ratio was 2.0.

Inspection of a time sequence of DNA histograms by an experienced observer enables a reasonable estimate to be made for the cell-cycle time, and two initial values of 12 and 14 hours were assumed. However, it is virtually impossible to make a reasonable guess for the CVs of the intermitotic phase times by mere inspection, no matter how experienced the observer. Thus, in order to test the optimiser rigorously each data set was analysed from four different starting sets of five initial points. Three of these were designed to be bad guesses, where the initial ranges of the CVs did not straddle the true values. The fourth was a good guess, where all the CV ranges straddled the true values. In addition, three criteria of fit – the simple sum of deviations, the sum of squared deviations and $\Sigma \chi^2$ – were used for each data set and for each of the initial starting conditions. Thus, a total of 60 computer runs were made.

The behaviour of the optimiser from a bad guess for the initial values of the population blocked in mid S phase (with the simple sum of deviations as the fit

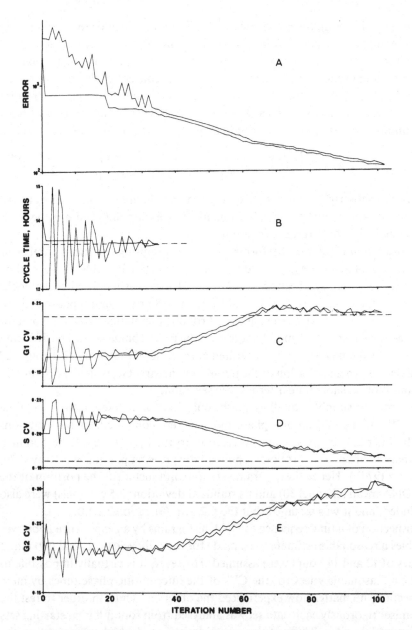

Figure 9.11. The error, panel A, associated with the best and worst 4-D co-ordinates of the four variables being optimised: t_c, CVt_{G1}, CVt_S and CVt_{G2+M}, which are shown in panels B, C, D and E respectively, all plotted against iteration number.

criterion) is shown in Figure 9.11, where the abscissa gives the iteration number. This illustration was chosen because this type of data set is one of the more difficult to analyse and required the longest computations: a total of 104 iterations were needed. Panel A gives the error associated with the best and worst

4-dimensional (4-D) points. Panels B, C, D and E (respectively) show the cycle time and the CVs of t_{G1}, t_S and t_{G2+M} associated with the greatest and least error. Note that the initial cycle-time range straddles the true value (dashed horizontal line), the initial ranges for the CV of t_{G1} and t_{G2+M} are well below the true values (dashed horizontal lines in panels C and E), and the initial range for the CV of t_S is well above the true value (dashed horizontal line, panel D).

Panel A of Figure 9.11 shows that the computation was divided into three distinct phases. The first occurred whilst the cycle time was converging rapidly towards the true value. It can be seen in panel B that the cycle-time value associated with the greatest error oscillated violently about its true and best values, exhibiting what may loosely be described as a damped harmonic wave. At iteration 35 the possible change in the cycle time that could occur between the best and worst values was so small (<0.1% of the best value) that this parameter was deemed to have converged, and the computation was automatically reduced from a 4- to a 3-dimensional problem. The predicted value was 13.28 hours, compared with the true value of 13.3 hours. During this phase the error on the worst 4-D set exhibited a declining saw-tooth pattern which tended to approach the error on the best 4-D set. It can be seen from panels C, D and E that little progress was made in the convergence of the remaining parameters, which tended to oscillate fairly incoherently. However, after convergence of the cycle time had occurred the second phase was entered, where the values of the remaining parameters "tracked" towards their true values in a regular manner. The error associated with the best and worst 4-D set also exhibited a regular decline during this second phase, which continued until the CV of t_{G1} overshot the true value. At this stage (iteration 62) the third phase was entered, where the error began to show a slower decline. The CV of t_{G1} now began to track back towards the true value after the overshoot, and the CV of t_S showed a slower and less regular convergence towards the true value. At iteration 74 the CV of t_{G2+M} overshot the true value. The possible change in the CV of t_{G1} that was allowed to occur between the best and worst values was so small by iteration 101 (<0.1% of the best value) that convergence was assumed by the program to have taken place. The predicted and true values were 23.3% and 23% respectively, and the computation was reduced from a 3- to a 2-dimensional problem. At the next iteration (102) the best and worst values for the CV of t_S converged at 16.9%, compared with the true value of 16%. The computation stopped at iteration 104, when the CV of t_{G2+M} converged to 30.8% compared with a true value of 28%. The total discrepancy between the actual and predicted number of cells was 3,657 out of 40,000 (4 histograms with 10,000 cells in each) for the worst 4-D set at the start of the procedure and 112 out of 40,000 for the best 4-D set at the end.

The overall results of the four stathmokinetic analyses are summarised in Figure 9.12, where the cell-cycle blocks were induced at 0.18, 0.38, 0.65 and 1.0 multiples of t_c, corresponding respectively to early/mid G1, early S, mid S and

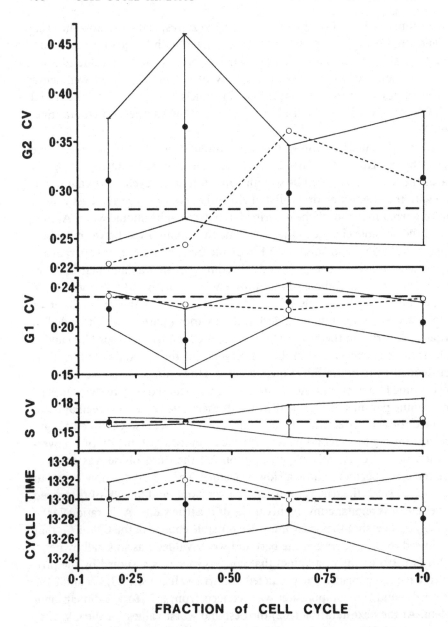

Figure 9.12. The overall results of the four stathmokinetic analyses, where the cell-cycle blocks were induced at 0.18, 0.38, 0.65 and 1.0 multiples of t_c, corresponding (respectively) to early/mid G1, early S, mid S and mitosis. The mean of all the predicted values for CVt_{G2+M}, CVt_{G1}, CVt_S and t_c are shown as the solid symbols, with their 95% confidence limits plotted against the position in the cell cycle at which the block was induced. The open symbols represent those values of the variables which gave the best result at each time point.

mitosis. The mean of all the predicted values for each of the parameters from all initial starting conditions is shown, plotted against the time relative to the cycle time at which the block occurred. The optimiser failed to converge in two runs, one for the early S and one for the mitotic block, and the mean values at these times were calculated from a total of 11 results as opposed to 12 for the early/mid G1 and mid S–phase blocks. Panels A, B and C (respectively) show the CVs for t_{G2+M}, t_{G1} and t_S, and panel D gives the cycle-time results. The solid symbols with error bars depicting the 95% confidence limits represent the mean values of the predicted results for the parameters from all runs in which convergence was achieved. The open symbols connected by the dotted lines represent the best-fit results, and the horizontal dashed lines give the true values.

The difference between the expected and predicted proportions of cells in each of the cell-cycle phases was investigated at each of the four times after the block commenced and for each of the four block conditions. Figure 9.13 shows

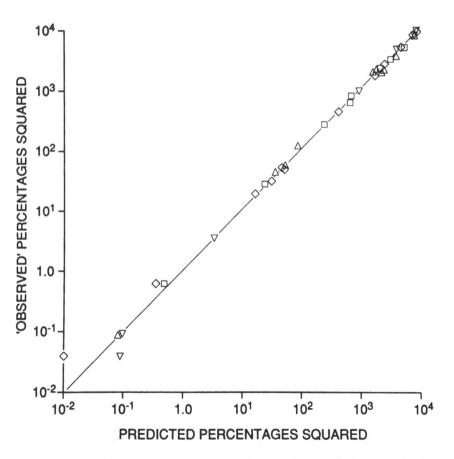

Figure 9.13. Known versus predicted proportions in the intermitotic phases for the four worst-fit stathmokinetic analyses. The data points from each analysis are distinguished by the different symbols, but there is no symbolic representation for different phases.

Table 38. *Results of synchrony progression analysis showing mean values of the parameters with 2 SE limits from 11 analyses*

Parameter[a]	Mean	2 × SE	Best fit	True values
CVt_S	14.9	±3.16	16.1	16.0
CVt_{G1}	22.3	±3.59	22.8	23.0
CVt_{G2+M}	31.3	±4.96	27.0	28.0
t_c	13.29	±0.018	13.3	13.3

[a] CVs are percentages; cycle time t_c in hours.

the square of the "observed" proportions plotted against the square of the predicted proportions on a log-log scale for the overall worst fits obtained for each of the four block conditions, independent of starting-set values and fit criterion used. Each point in Figure 9.13 should lie on the line of unit slope; this type of display was chosen so that any discrepancies would be magnified. However, in spite of this unflattering method of presenting the information, it can be seen that there is very little discrepancy between predicted and observed values from unity to 10^4 on the scales shown. This represents a range from 1% to 100% in a given cell-cycle phase. It was only values below 1.0% that gave rise to marked divergence from unit slope, and this result was obtained with the worst fits and sets of results.

The overall results for the analysis of the synchrony progression simulation are given in Table 38, which shows the mean values of the CVs of t_{G1}, t_{G2+M} and t_S and the mean value of the cell-cycle time, together with their 95% confidence limits. These results were calculated from a total of 11 analyses, as the optimiser failed to converge in one run. Also included are the best overall results for the computed parameters and their true values.

The mean values of the proportions of cells in each cell-cycle phase were calculated from the 11 successful runs, and these results are shown in Figure 9.14. The curve represents the true values plotted against time (after synchronisation), and the error bars on the means of the predicted data (where these are greater than the diameters of the symbols) represent the 95% confidence limits.

This optimising, or search-minimisation, technique is similar in principle to the Nelder–Mead simplex method (1965). It cannot be guaranteed to converge to a unique solution, particularly when the error surface is likely to be irregular. However, as the second derivatives of the various mathematical functions of the model cannot be obtained analytically, it is probably one of the better types of procedures for obtaining a solution to this class of multidimensional problem.

The inability to converge to a unique solution is illustrated by the results of the analyses, and the particular solution for any one run was somewhat

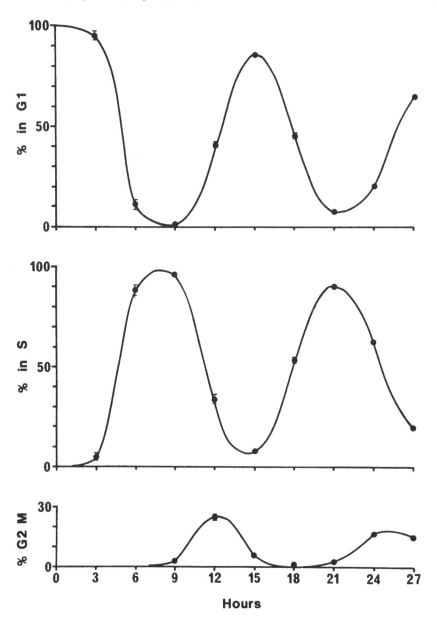

Figure 9.14. Proportion in G1, S and G2+M from the synchrony progression analysis. The curves represent the known and the symbols represent the predicted proportions versus time after synchronisation in early G1. The error bars, where these are greater than the diameter of the symbol, depict the 95% confidence limits.

dependent on the initial guesses made for the variables. No dependence was found on the criterion of fit, for example $\sum d$, $\sum d^2$ or $\sum \chi^2$. However, relatively small deviations of some variables from their true values will give rise to large discrepancies between the model-predicted and simulated (or, in the real world, experimentally determined) numbers of cells per channel of the DNA histograms. Such parameters are said to be well "conditioned", and in this study the conditioning was in the order $t_c > CVt_S > CVt_{G1} > CVt_{G2+M}$ from best to worst. Also, t_c was always first and CVt_{G2+M} was always last to converge. This is to be expected, as the cell-cycle time will always be better conditioned than the CVs of the phase times. The conditioning hierarchy of the CVs was also to be expected from this data set. The CV of t_S was 16%, the smallest of the three, and the duration of S phase was greater than either t_{G1} or t_{G2+M}. The CV of t_{G2+M} was greater than either of the other phase time CVs and the duration was shorter. Thus, we would expect the conditioning of CVt_{G2+M} to be worse than that of either t_S or t_{G1}. It is possible, therefore, to make the generalisation that in cell-cycle analysis the degree of conditioning is directly proportional to the duration and indirectly proportional to the phase time CV.

This technique has been found to be generally reliable and robust, and it can deal with bad initial guesses (see Figure 9.11). In 3 of the 60 runs, however, the optimiser was unable to achieve convergence, and in each of these failures a "saddle" error contour pattern was located. This problem occurs when the worst point can only move into an error domain that is always greater than the second worst point, and it invariably indicates a complicated error domain with local minima scattered throughout. The optimiser has a routine that enables a re-start to be made once a saddle error contour has been located, and in 6 of the 57 successful runs a re-start was made automatically and convergence was subsequently achieved. However, as the number of variables being computed increases, the greater will be the chance of locating an irregular error domain and the greater will be the difficulty encountered in "climbing" out of a local minimum to continue the search. When failure to converge occurs after a saddle error contour has been located, there is little alternative but to re-start the optimiser from a completely different set of initial values.

The problem with analysis of a simulated data set is that we know the answer we are looking for, whereas with a real data set we obviously do not. Furthermore, the simulated histograms were perfect, containing neither experimental artefact nor statistical fluctuation; this contrasts sharply with real data. In spite of this advantage, in these analyses it was not possible (within the constraining limits) to "get back" precisely to the true values of all the parameters from any one starting set. As mentioned previously, the cycle time was the best-conditioned variable and this was followed by the CV of t_S. It can be seen in Figure 9.12 and Table 38 that while these parameters were reliably predicted there was proportionally greater variability for the predictions of the CVs of t_{G1} and t_{G2+M}, particularly in the stathmokinetic analyses (see Figure 9.12). The CV prediction for t_{G1} was the worst (as expected) when the block was at the G1/S interface, and this caused considerable spread in the CV of t_{G2+M}. The predicted mean value for CVt_{G2+M} was also well above the true value (Figure 9.12), and with a large spread, when the block was at the end of the cycle; this also was expected. These factors highlight the necessity of starting the optimiser (particularly for real data) from a number of different initial sets, and it is possible that an n-dimensional computation will need $n+1$ starting sets. However, if after three runs one of the parameters (e.g. t_c) converges to approximately the same solution each time then it would probably be valid to fix the value at the mean of the three and exclude it from subsequent runs, thus reducing the number of variables and computing time.

From a purely biological standpoint there is no virtue in obtaining values for the cell-cycle time and phase-time CVs to within the limits shown here if no additional biological information will be forthcoming. The biologist would wish to know the proportions of cells in the various phases of the cycle from this class of data, and furthermore how a perturbing event (e.g. cancer-therapy agents) might change those proportions. The data illustrated in Figure 9.13 show that significant discrepancies between known and predicted proportions occurred only at or below the 1% level, even for the worst predictions of the computed variables. Figure 9.14 also shows that the differences are small and of the order of at most a few percentages. This is well within the biological "noise level" that we all experience.

9.5   FPI analysis

The various equations shown in previous sections may look somewhat fearsome to the uninitiated at first sight; however, they are not really very difficult to understand and with a little perseverence most people can appreciate them. Unfortunately, the majority of commercial computer packages do not contain programs that can analyse a time sequence of histograms as shown in Figure 9.3. There is, however, one very elegant and simple analysis procedure by Zeitz and Nicolini (1978) which can be carried out with only a piece of graph paper and a pocket calculator. Because of its simplicity it is readily available to all users and hence deserves particular attention.

Let us consider again what is happening to the biology in the data set shown in Figure 9.3. A well-synchronised G1 population is progressing through G1, then doubling its DNA content during S and dividing at the end of the G2+M phase. This is represented longitudinally in time in Figure 9.15. We can arbitrarily consider a particular DNA level between G1 and G2+M, and at any time point there will be a flux into and a flux out of this compartment. For simplicity let us consider the mid S phase in Figure 9.3. The input and output fluxes at this DNA level will both be zero until some of the G1 synchronised cells have increased their DNA sufficiently to be scored with 3C DNA content. The input

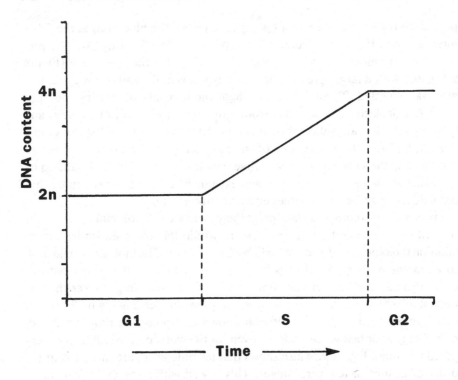

Figure 9.15. Longitudinal representation of the cell cycle.

flux will rise faster than the output flux until half the population has passed through this DNA level. Thereafter, the output flux will exceed the input. Thus, the fraction of the whole population with this DNA level will rise and then fall as the synchronised cohort of cells passes through this compartment. A second rise then fall will occur during the second cycle after division has taken place. There will, however, be some damping of the amplitude of the wave owing to de-synchronisation. The frequency, or periodicity, of the wave through this compartment will be that of the cell-cycle time.

In practice one should not consider a single channel of the DNA histogram (DNA level), but rather one or two on either side of the arbitrarily fixed level in order to allow for statistical fluctuations in the data. Once these limits have been chosen the calculation merely involves dividing the number of cells within the limits by the total number of cells in the population. Figure 9.16 shows the fraction of the whole population in (FPI) the selected (here, mid S) phase from Figure 9.3 plotted against time. The cell-cycle time was approximately 13.0 hours compared with 13.1 hours from the mathematical model. By selecting different DNA levels (e.g. FPI in G1 or G2+M) it is possible to obtain estimates of the intermitotic times, which were 5.0, 7.0 and 1.0 hours for G1, S and G2+M respectively, very close to the computer model results.

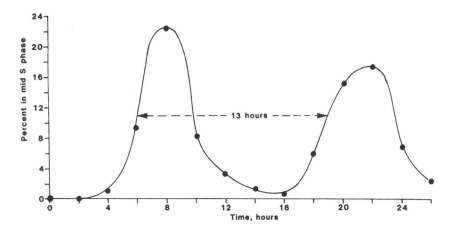

Figure 9.16. Fraction of the whole population in a small time domain in mid S phase. The periodicity of the curve is that of the cell cycle.

9.6 Bromodeoxyuridine

Bromodeoxyuridine (BrdUrd) is a thymine analogue which is incorporated into DNA during synthesis. Two methods involving BrdUrd uptake are available for studying kinetic changes, where the steady state of growth remains unperturbed. The fluorescence from the bisbenzimidazole dyes (e.g. Hoechst 33558 and 33342) is quenched by the presence of BrdUrd in DNA, as shown by Latt (1977); Figure 9.17 is redrawn from that publication. Mixtures of the polynucleotides poly(dA.BrdUrd) and poly(dA.dT), where the molar proportion of the former was varied from 0 to 1, were assayed fluorimetrically after addition of Hoechst 33258 and ethidium bromide. With an increasing fraction of poly(dA.BrdUrd) there was a progressive decrease in Hoechst 33258 fluorescence, reaching a limit at about 15% of that of the poly(dA.dA) control. This decrease in Hoechst fluorescence with increasing poly(dA.BrdUrd) concentration was accompanied by a very small increase in ethidium bromide fluorescence. Following this observation, Latt, George and Gray (1977) were able to show that CHO cells incubated with BrdUrd for 24 hours and stained with Hoechst 33258 could be distinguished from similarly stained control cells not treated with BrdUrd. These data are reproduced in Figure 9.18, which shows the fluorescence profiles of the BrdUrd-treated and untreated cells together with a 1:4 mixture of the two.

Beck (1981), Bohmer (1979), Hieber, Beck and Lucke-Huhle (1981) and Nusse (1981) have all used BrdUrd quenching of Hoechst 33258 fluorescence in cell kinetic studies by following time courses during continuous labelling. However, the method can have resolution problems, as quenching is directly proportional to the quantity of BrdUrd incorporated. It can be difficult to obtain

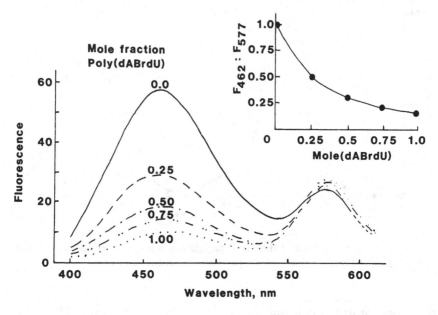

Figure 9.17. Quenching of Hoechst 33258 fluorescence by increasing molar ratios of poly(dA.BrdUrd). Ethidium bromide was also included in the sample, and there was a slight increase in EB fluorescence with increasing proportion of poly(dA.BrdUrd). The insert shows the ratio of fluorescence at 462 nm (Hoechst) to that at 577 nm (EB) versus poly(dA.BrdUrd) concentration which represents poly(dA.BrdUrd) content. Redrawn from Latt (1977).

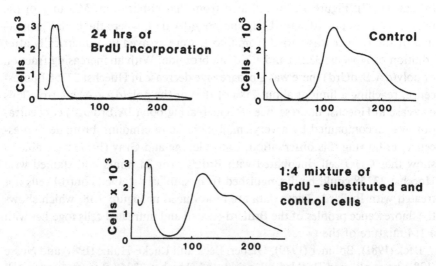

Figure 9.18. Quenching of Hoechst 33258 flourescence in tissue culture cells. Top left, cells exposed to BrdUrd for 24 hours; top right, control cells. The bottom panel shows a 1:4 mixture of BrdUrd-treated and untreated cells.

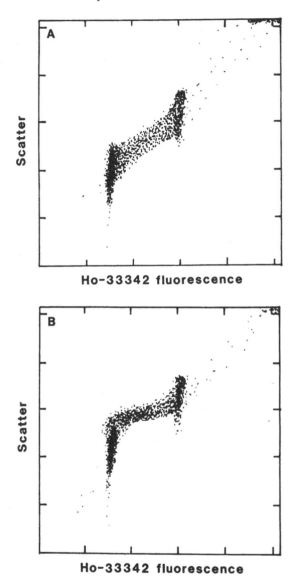

Figure 9.19. Flash labelling of tissue culture cells with BrdUrd for two hours, where panels A and B show the control and treated cells respectively. Some quenching of Hoechst 33342 fluorescence in the S-phase cells can be seen in panel B. The quenching "moves" these cells to the left and they then tend to overlie each other; some overlapping with G1 cells, which have not incorporated, occurred.

sufficient uptake of BrdUrd in flash labelling experiments to monitor the S-phase fraction independently of G1 and G2+M, since quenched S-phase DNA overlaps the G1 component. This is illustrated in Figure 9.19, where 90° light scatter is plotted on the ordinate versus Hoechst-33342 fluorescence on the abscissa as dot-plots for both control and BrdUrd-treated cells (5 μg ml⁻¹ BrdUrd for 2 hours), panels A and B respectively. The fluorescence quenching of the S-phase fraction in the BrdUrd-treated cells is clearly apparent, and many of these would overlap the G1 peak (see panel B). The separation we obtained was not regarded as being sufficient to enable us to use this method for BrdUrd pulse–exposure experiments where the S-phase fraction was to be followed with time. However, this method has been used for continuous labelling experiments, which will be described in Section 9.6.2.

9.6.1 Pulse labelling

The problems encountered with BrdUrd quenching have been overcome by the second method using this analogue due to Gratzner et al. (1975), who raised a polyclonal antiserum to the DNA–BrdUrd complex in rabbits. Some of the original data obtained after BrdUrd incorporation and immunoperoxidase staining are reproduced in Figure 9.20, which shows patchy staining in a nucleus and two chromosomes, one of which exhibits sister chromatid exchange. The first use of this technique in flow cytometry was reported by Gratzner and Leif (1981), where dual parameter data of light-scatter signals versus BrdUrd fluorescence were used to identify the labelled-cell fraction after a 60-minute exposure to 10 μM BrdUrd. Gratzner (1982) produced monoclonal antibodies to both BrdUrd and IdUrd as a means of detecting DNA replication, and Dolbeare et al. (1983) extended this method to measure total cellular DNA simultaneously with BrdUrd incorporation. Cells were pulse labelled with the analogue, treated with a nuclear isolation buffer, fixed and then subjected to an hydrolysis step to denature double-stranded DNA to single strands (the anti-BrdUrd antibody recognises BrdUrd only in single strands). The nuclei were then incubated with the antibody which was subsequently tagged with fluorescein (green), and total DNA was stained with propidium iodide (red). The redrawn data from Dolbeare et al. (1983) are shown in Figure 9.21. The

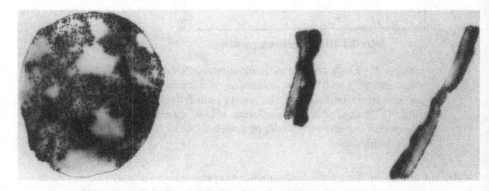

Figure 9.20. Reproductions of some of Gratzner's original immunoperoxidase data (Gratzner et al. 1975) showing incorporation of BrdUrd into a nucleus and two chromosomes, one of which (right) exhibits sister chromatid exchange.

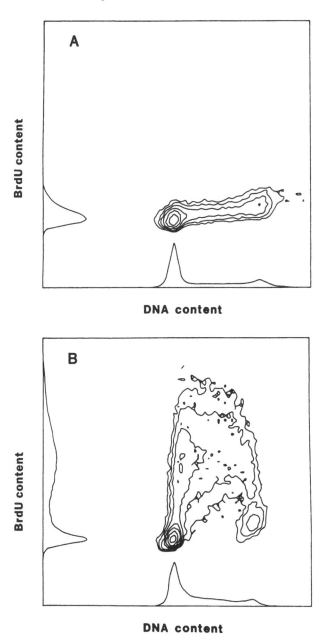

Figure 9.21. Bivariate analysis of BrdUrd incorporation (green fluorescein fluorescence) on the ordinate versus total DNA (red propidium iodide fluorescence) on the abscissa for tissue culture cells. Controls in panel A; BrdUrd incorporation in panel B. Redrawn from Dolbeare et al. (1983).

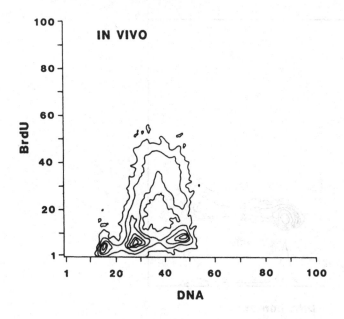

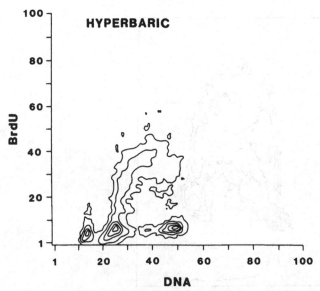

Figure 9.22. An in vivo tumour showing BrdUrd incorporation after flash labelling (top panel) and the following hyperbaric oxygen treatment (bottom panel). Both data sets show some cells with S-phase DNA content which have not incorporated BrdUrd. However, this component is reduced in the tumour treated with hyperbaric oxygen, suggesting that these S-phase cells are nutrient limited. Redrawn from Wilson et al. (1985).

BrdUrd-associated fluorescence (green) is plotted on the ordinate versus DNA fluorescence (red) on the abscissa as a contour map, where panels A and B (respectively) show the control and the BrdUrd-incorporation data after flash labelling. The horseshoe pattern in panel B is characteristic, with G1 and G2+M cells showing no BrdUrd-associated fluorescence and the S-phase cells exhibiting differing degrees of incorporation; this is related to the rates of uptake at different stages of DNA synthesis. Dean et al. (1984) have shown how the rate of DNA synthesis during S phase can be calculated from data such as those in Figure 9.21.

Similar data sets are shown in Figure 9.22, where an in vivo mouse tumour was analysed both before and after hyperbaric oxygen treatment (Wilson et al. 1985). In both panels there is a component with S-phase DNA content which has not been stained by the anti-DNA–BrdUrd antibody, indicating that some cells with an S-phase DNA content are not synthesising DNA. However, after hyperbaric oxygen this component is decreased, suggesting that some tumour cells are nutrient limited. A time-course experiment using this technique in a mouse tumour is shown in Figure 9.23 (Begg et al. 1985). Immediately after the BrdUrd pulse the data are qualitatively similar to those in Figures 9.21 and 9.22. However, with increasing time between BrdUrd administration and sampling of the population, there is movement of the labelled cohort through S phase and then G2 and division to appear in the G1 region of the 2-dimensional map with a halving of their green fluorescence intensity. It can also be seen that cells with a G1 DNA content at the time BrdUrd was administered track through the S-phase domain with no green fluorescence. It is obvious from the data shown in Figure 9.23 that we can now track the behaviour of the two different subsets with time, namely cells that were synthesising DNA when BrdUrd was administered and those that were not. Calculation of the proportions in each cell-cycle phase can now be carried out and these data can be used to estimate the cell-cycle time and intermitotic phase times, as we will now show.

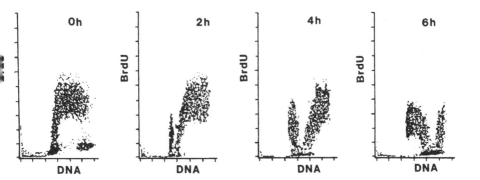

Figure 9.23. BrdUrd pulse–chase experiment where the in vivo tumour was followed at two hourly intervals up to six hours. Redrawn from Begg et al. (1985).

This method has been used in our laboratories to study the effects of the c-*myc* protein on the cell cycle and intermitotic phase-time durations. The c-*myc* gene is activated very early in the cell cycle after stimulation of quiescent cells. This points to some form of control mechanism acting at the G_0-to-G1 transition, with elevation being required not only to facilitate entry of cells into the cell cycle but also to maintain cells in a potentially active proliferating state. Some evidence suggests that p62^{c-myc} participates in DNA synthesis (Studzinski et al. 1986). However, cells are extraordinarily efficient and logical in performing their various functions, and there is no reason for the c-*myc* gene to be activated early in G1 to produce mRNA for p62^{c-myc} with both being turned over with half-times of 20–30 minutes (Hann, Thompson and Eisenman 1985; Rabbitts et al. 1985) if the protein is required only in S phase.

Because of these considerations, Jon Karn of the Laboratory of Molecular Biology, Cambridge, constructed a series of retroviral vectors containing the c-*myc* gene in various configurations to investigate p62^{c-myc} cell-cycle dependency using the BrdUrd antibody technique. These retroviral constructs were composed of SV-40 promotor regions (SV0), c-*myc* long terminal repeat sequences (LTR), a c-*myc* "mini-gene" containing the protein-coding exon sequences without the introns, and the neomycin resistance gene (NEO) for selection purposes. A number of retroviruses were constructed and the structures of three of these, VSN-2, NSM-4 and MSN-7, are shown in Figure 9.24. All three viruses contained SV0 (SV-40 promotor), the NEO gene and *myc* LTRs, but

VSN-2

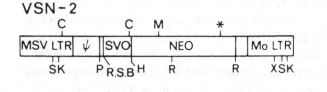

NSM-7

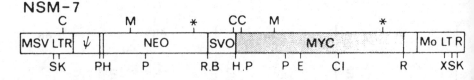

MSN-4

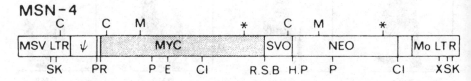

Figure 9.24. Structures of the retroviral vectors carrying the human c-*myc* gene with introns sliced out (stippled), drawn in the form of proviral DNA constructs cloned in *E. coli* plasmids.

only NSM-4 and MSN-7 contained the *myc* gene. In NSM-4 the *myc* gene was under the control of SV0, but in MSN-7 it was under control of its own LTR. After cells were infected with these viruses there was no enhanced production of p62$^{c\text{-}myc}$ in the VSN-2 cell line, some enhancement in the NSM-4 cells and considerable enhancement in the MSN-7 cells, which was the expected result (Karn et al. 1989). The growth rates of the three cell lines showed progressive increases with p62$^{c\text{-}myc}$ amplification, with VSN < NSM < MSN (which I'll abbreviate to V-, N- and M-cells).

The cell-cycle kinetics were determined by following the changes in BrdUrd signals versus total DNA at hourly intervals after flash labelling. Selected data sets are shown in Figure 9.25 for the three cell lines. The left, middle and right columns show the data from the V-, N- and M-cells (respectively), and the rows, from top to bottom, show the data obtained at 0, 4, 8 and 12 hours after

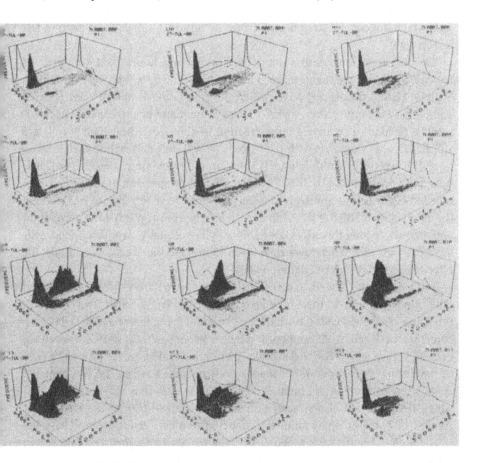

Figure 9.25. Bivariate perspective histograms of DNA versus BrdUrd for V-, N- and M-cells in the columns at 0, 4, 8 and 12 hours after pulse labelling in the rows. The origins are on the left, with DNA displayed from top left to bottom right and BrdUrd incorporation displayed from bottom left to top right.

labelling. The origins in these 3-dimensional displays are on the left of each panel, where BrdUrd-associated green fluorescence (520RF AREA) is scored obliquely from bottom left to top right and total DNA (630RF AREA) is scored from top left to bottom right. Frequency is scored on the vertical axis. The large spikes in the top three rows, at the time-zero data point for each of the three cell lines, represent unlabelled G1 cells and are centered at channel 300 on the DNA axis. The S-phase fractions are represented by the characteristic U-shaped patterns and the G2+M cells, which are also unlabelled, are represented by the small spikes centered at channel 600 on the DNA axis. The second row shows the data at 4 hours. In each panel the most obvious change is that some cells which were in late S phase when the BrdUrd was added have progressed through G2+M, split into two and in so doing have halved their BrdUrd content, now appearing with G1 DNA content at channel 300 on the DNA axis (520RF AREA). These are the G1 cells with the positive green signals which appear to "stream out" from the large G1 spikes from bottom left to top right.

The next row down shows the data obtained at 8 hours, and a number of further changes have taken place. First, most of the cells that were initially in S phase (BrdUrd positive) have passed through G2+M, halved their BrdUrd content and are now scored with G1 DNA content and positive green (BrdUrd) signals. Second, a relatively large fraction of the BrdUrd-labelled V-cells are still in G2+M compared with the M-cells (row 3, left and right panels respectively). The N-cells (middle panel) show an intermediate pattern. Third, it is now quite obvious that there has been significant progression of the initially unlabelled G1 cells that are BrdUrd negative. These have moved out of G1 along the DNA axis (630RF AREA) in the direction from top left to bottom right and are progressing towards G2+M. Finally, it is equally clear that the M-cells have progressed further towards G2+M than have the V-cells, and again the N-cells occupy an intermediate position.

By 12 hours the M-cells (right panel, bottom row) are exhibiting a pattern qualitatively similar to their time-zero counterpart (right panel, top row), but the BrdUrd-labelled cells have half their original green fluorescence signals. In contrast, the V-cells at 12 hours (left panel, bottom row) still show some cells in G2+M with positive BrdUrd signals (these are situated on the rear right wall of the display as the spike), but the majority of these have a G1 DNA content. Also, some of the initially unlabelled G1 cells are still in S phase and G2+M, where the latter are clearly seen as a spike at channel 600 on the DNA axis. Again, the N-cells display an intermediate pattern between the V- and M-cells.

The data in Figure 9.25, together with the remaining majority of data sets in the experiment, were analysed by placing gates around both the labelled and unlabelled G1, S and G2+M regions to obtain the proportions in each of these compartments. Figure 9.26 illustrates this procedure. Panel A shows the control time-zero data for the V-cells and panel B shows the zones used in the

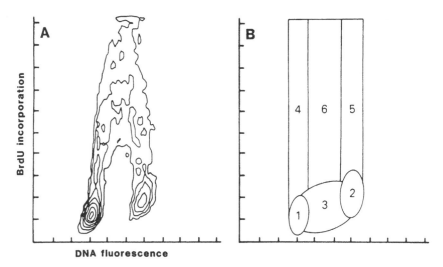

Figure 9.26. Panel A shows BrdUrd uptake into V-cells at time zero, where the characteristic incorporation horseshoe is readily evident. Panel B shows the gates used to assess the unlabelled proportions in G1, S and G2+M (zones 1, 3 and 2 respectively) and the labelled proportions in G1, S and G2+M (zones 4, 6 and 5 respectively).

gating analysis procedure, where 1, 3 and 2 correspond to unlabelled G1, S and G2+M cells (respectively) and where 4, 6 and 5 correspond to labelled G1, S and G2+M cells. The data for the unlabelled population are shown in Figure 9.27, where the various proportions are plotted as the points on the ordinates versus time on the abscissae. The cell-cycle phases G1, S and G2+M are arranged in the columns for V-, N- and M-cells in the rows.

Figure 9.28 shows the comparable data for the labelled cells, that is, those initially in S phase. The curves were calculated with the model described in Section 9.3. The durations of G1, S and G2+M with their 95% confidence limits are shown in Figure 9.29, where it can be seen that the durations of S phase and G2+M were not significantly different in the three cell types. However, there was a progressive reduction in the duration of G1 from V- to N- to M-cells. Further experiments were carried out where the three cell types were BrdUrd pulse–chased into serum-free medium. Again, the durations of S phase and of G2+M were not altered, but there was a progressive increase in the length of G1 which was greater in V- than in N- than in M-cells. These data not only add support to the hypothesis that $p62^{c\text{-}myc}$ is involved in cell-cycle regulation, but also suggest very strongly that this control is exerted in G1.

The BrdUrd antibody technique is potentially of very great importance in cell biology. However, it can be a little tricky, and a considerable amount of work has been carried out on the stoichiometry and sensitivity (Dolbeare et al. 1985) and the role of the denaturation step (Moran et al. 1985). Beisker, Dolbeare and

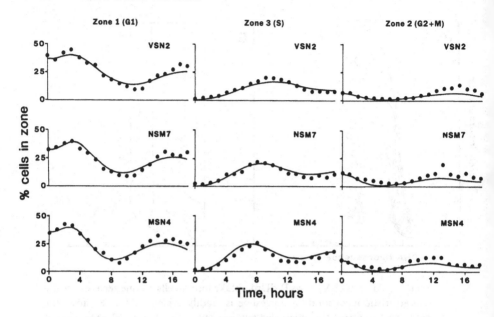

Figure 9.27. The proportions defined in zones 1, 3 and 2 of Figure 9.25, unlabelled cells, plotted as the points on the ordinates versus time on the abscissae. The cell-cycle phases G1, S and G2+M are arranged in the columns for
V-, N- and M-cells in the rows. The curves were calculated from the Hartmann-
Pederson type of computer model (see Section 9.3).

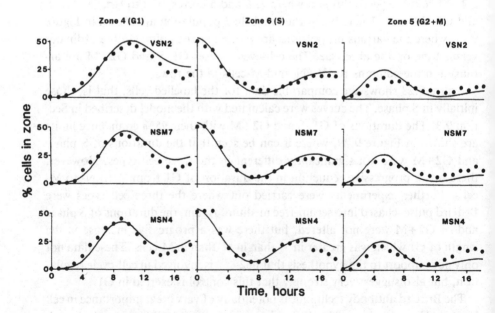

Figure 9.28. Similar display to Figure 9.27 for zones 4, 6 and 5 (G1, S and
G2+M respectively) of Figure 9.25, representing the labelled cells that initially
were in S phase.

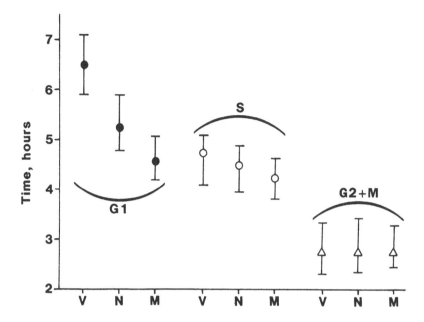

Figure 9.29. Durations of G1, S and G2+M with their 95% confidence limits. The durations of S phase and G2+M were not significantly different in the three cell types. However, there was a progressive reduction in the duration of G1 from V- to N- to M-cells.

Gray (1987) have also improved the immunocytochemical procedures to obtain increased detection sensitivity. This is not a technique to be undertaken lightly, and each cell type should be investigated in pilot studies before embarking on full-blown experiments. If you are really serious about using this method then you should acquire a copy of the November 1985 issue of *Cytometry* (volume 6), which was entirely devoted to the technique, and do some specialized reading. This issue has also been published as a book, edited by Gray and Mayall (1985).

9.6.2   Continuous labelling

Shortly after the work of Latt (1977) and Latt et al. (1977) introduced the Hoechst-BrdUrd quenching technique it was used by Beck (1981), Bohmer (1979), Hieber et al. (1981) and Nusse (1981) to study various aspects of cell-cycle perturbation. Some of these results, from control unperturbed Ehrlich ascites tumour cells growing in vitro, are reproduced in Figure 9.30 (Nusse 1981). The cells were continuously exposed to BrdUrd from time zero, sampled at intervals, stained with Hoechst 33258 and analysed. The top left panel shows the time-zero DNA histogram with no incorporation. At two hours, $t = 2h$, the G2+M peak has decreased in size as these cells, designated G2′, now reside in the G1 peak. Any late S–phase cells incorporating BrdUrd in the first two hours after labelling have undergone some quenching of Hoechst 33258 fluorescence.

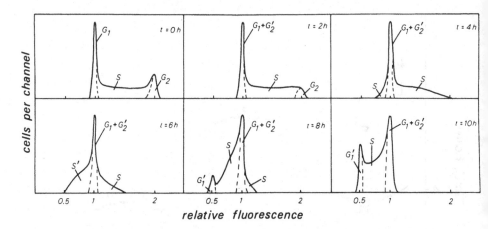

cells per channel

relative fluorescence

Figure 9.30. Time course of Hoechst 33258–stained DNA histograms at 2-hour intervals after addition of BrdUrd. G1, S and G2 denote cells that have not incorporated the analogue. G1', S' and G2' denote cells that have incorporated the analogue and exhibit fluorescence quenching. Redrawn from Nusse (1981).

On subsequent entry to G2 followed by mitosis and division, they will halve their DNA containing BrdUrd (which is quenched), and as G1 cells they will be scored just to the left of the non-quenched G1 peak. This is not apparent at the 2-hour point $t = 2h$, but it is apparent at 4 hours ($t = 4h$), where these quenched G1 cells are designated S'. With increasing time after BrdUrd addition the S-phase cells will incorporate relatively more BrdUrd and will suffer relatively greater fluorescence quenching; hence the non-quenched S-phase compartment, designated S, will decrease and the quenched G1 compartment, S', will increase (panel $t = 6h$). Any cells that were in G1 when BrdUrd was added will traverse the whole of S phase during the first cycle, exhibit maximum quenching and, following division, will appear as the population designated G1' which is apparent at 8 hours, panel $t = 8h$, and increases further at 10 hours, panel $t = 10h$.

The numbers of cells in these various subpopulations were calculated and plotted against time in Figure 9.31 where, as can be seen, the number of cells in S' and G1' increased linearly with time. The duration of G2+M is the interval between time zero and the time at which the S' line intersects that of G2, and the S-phase duration is from this point to the intersection of the G1' line with that of G2, as indicated on the figure. The total duration of the cell cycle was obtained from the increase in cell numbers with time, hence the duration of G1 was obtained by difference.

This method was very satisfactory for investigating the effects of irradiation on cell-cycle delay during the first cell cycle after BrdUrd was introduced into the system (Hieber et al. 1981; Nusse 1981) but it does have resolution problems for second and subsequent cycles. By inspecting Figure 9.30 we can see that during the next cycle the fluorescence will be further quenched, shifting the data further to the left and thus reducing the ability to delineate the various subsets.

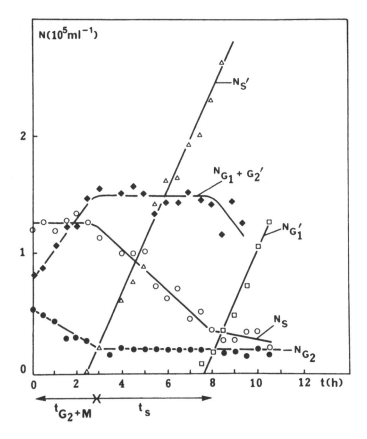

Figure 9.31. The numbers of cells in the various compartments versus time after BrdUrd addition. Redrawn from Nusse (1981).

9.7 Human tumour kinetics

It was pointed out in Section 8.1 that the cell-cycle time is invariably less than the observed doubling time of tumours growing in vivo. Mendelsohn (1962) suggested that only a portion of the tumour cells were within the replicative cycle to explain this discrepancy, and coined the term "growth fraction". However, it soon became apparent that the ^{3}H-thymidine labelling index tended to be too high for the very low growth fractions that would have to be postulated for the discrepancies in a number of tumours. This gave rise to the concept of cell loss from tumours, which was compatible with observations of necrotic regions seen in most tumours.

A major contribution to this area of cell kinetics was made by Steel and colleagues (Steel 1968, 1973; Steel, Adams and Barrett 1966) at the Institute of Cancer Research. Using age distribution theory in conjunction with cell-cycle duration estimates, it was shown that the growth fraction GF was equal to the expression $\alpha - 1$, where α is the number of proliferating cells produced per proliferating cell per cell-cycle time. It was also shown that the potential doubling time of a tumour, T_{pot}, defined as the time it would take for that tumour to double its volume if there were no cell loss, could be calculated from the duration of the DNA synthesis phase (S phase) t_S, the labelling index LI, and a constant λ which is dependent on α and the position of S phase in the cell cycle. This relationship is

$$T_{pot} = \frac{\lambda \times t_S}{LI}.$$

The relationship between the labelling index and potential doubling time when the durations of t_{G1} and t_S are known was shown in Section 9.4.1, and the potential doubling time T_{pot} is related to the rate constant for cell production, K_p, by the following equation:

$$K_p = \frac{\log_e(2)}{T_{pot}}.$$

The rate constant for cell production can now be expressed in terms of λ, t_S and LI by substituting the right-hand side of the first equation for T_{pot} in the second to give

$$K_p = \frac{\log_e(2) \times LI}{\lambda \times t_S}.$$

The growth-rate constant K of a tumour is equal to $K_p - K_L$, where K_L is the rate constant for cell loss. Hence the overall growth of a tumour is given by the expression

$$y = y_0 \exp\left[\left(\frac{K \log_e(2)}{T_d}\right)t\right] = y_0 \exp\left[\left(\frac{(K_p - K_L)\log_e(2)}{T_d}\right)t\right],$$

where t is time and T_d is the observed doubling time of the tumour. The rate constant for cell loss K_L can thus be obtained by comparing the observed doubling time T_d and the potential doubling time T_{pot}. A parameter called the cell loss factor (CLF) can now be calculated as the ratio of K_L to K_p, which expresses the rate of cell loss in terms of the rate of cell production. In experimental animal tumours the cell loss factor is usually within the range 0.4 to 0.8, but in the relatively few human tumours in which this has been measured values of 0.85 to 0.95 have been obtained. This means that the rate at which cells are dying and are lost within a tumour can be up to about 95% of the rate at which cells are being produced, and the net result is that the overall growth rate is very "slow" in comparison with the rate at which proliferation is taking place.

Non-surgical cancer therapy (radiotherapy or chemotherapy) is usually fractionated, but the optimum interval between fractions is not known. However, radiotherapy tends to be delivered 3 or 5 times per week while chemotherapy tends to be delivered every 3 or 4 weeks. Moreover, all tumours of a given type tend to be treated in a similar manner, taking no account of any possible variation in growth rate between tumours in the group.

The primary determinant of tumour growth rate is the rate of cell production, but there was no realistic means of assessing this until Begg et al. (1985) produced a very elegant method which I consider to be potentially of very great importance in cancer therapy. That importance lies in the simple fact that for the very first time we are able to measure something in a tumour which has direct relevance to the speed at which the tumour is growing, and thus prescribe treatment specifically for that tumour. Because of its potential importance I'll describe it in some detail, though the reader is encouraged to read the original paper (Begg et al. 1985) and the follow-up correspondence (White and Meistrich 1986).

Begg et al. (1985) suggested that a single dose of bromodeoxyuridine should be administered to patients and then, at a defined interval later (about 4 hours), a single biopsy should be taken from the tumour. Part of this biopsy would be used for the normal histological diagnosis and part for flow cytometric analysis. During the interval between administration and biopsy, some of the initially labelled S-phase cells will have halved their BrdUrd content at mitosis and will have divided to appear in G1. Let us suppose for simplicity that 50% of the cells have halved their BrdUrd content and entered G1 4 hours after administration. It's not too difficult to appreciate that the remaining 50% will also take about 4 hours to halve their BrdUrd content and enter G1, giving a total S-phase duration of about 8 hours. Similarly, if after 4 hours only 25% of labelled cells have halved their BrdUrd content and entered G1 then there will be three more cohorts of 25% of the total, each of which will require about 4 hours to complete DNA synthesis. Thus, to a first approximation, the S-phase duration will be about 16 hours. Furthermore, the labelled fraction can be estimated, and assuming a value for λ of about 0.9 it is possible to obtain the potential doubling time.

In principle it really is as simple and as elegant as that, but in practice there are a number of problems. For instance, it must be remembered that every two BrdUrd positive cells in the G1 peak after division started off as a single cell when the label was administered; this has to be taken into account in the calculations. Additionally, there is no way of knowing the optimum timing of the biopsy before it is taken. Ideally it should be taken when 50% of the cells have exited G1, and occasionally – when there is a very long S-phase duration – the 4–6 hour interval is too short. If either too few or too many cells have exited S phase then uncertainty in estimating the S-phase duration increases. Also, many human tumours contain large volumes of necrosis, so the results obtained will to some extent be sampling dependent. Hence, so-called representative biopsies from a number of different areas should be obtained.

Nevertheless, in spite of these potential problems, the technique is very important and is now being used in Medical Research Council clinical trials at

Mount Vernon Hospital, Northwood. Patients with high rate constants for cell production (K_p calculated from T_{pot}) are assigned to hyper-fractionated radiotherapy schedules (multiple fractions per day), those with low values of K_p to conventional single daily fraction schedules. The difference between this clinical trial and most others is that the two arms of this study are designed around a highly relevant biological parameter, the cell production rate, which is actually being measured within the tumour; flow cytometry is the only means of doing this.

9.8   Acridine orange

Acridine orange has been used for many years to obtain a measure of DNA and RNA simultaneously (Darzynkiewicz et al. 1975, 1977a,b,c; Traganos et al. 1977). Bromodeoxyuridine incorporation into DNA also suppresses acridine orange DNA fluorescence (Darzynkiewicz et al. 1978). As this analogue is not incorporated into RNA, it was possible to identify cells "in transit" from G_0 to G1A (G1T) on their RNA content, where their DNA fluorescence was not suppressed. These data are shown in Figure 9.32, with DNA plotted on the ordinate versus RNA on the abscissa. Panel A shows G1Q (G_0) non-stimulated lymphocytes. In panel B the phyrohaemagglutinin (PHA)–stimulated cells had been exposed to BrdUrd for approximately one cell cycle, thus all cells within

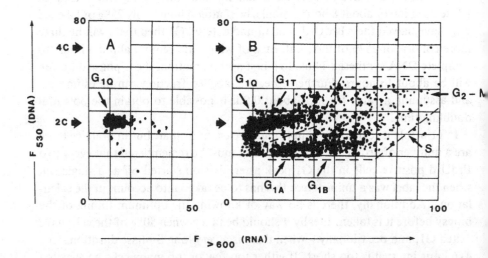

Figure 9.32. BrdUrd quenching of acridine orange green fluorescence (DNA) in stimulated lymphocytes, where green (DNA) fluorescence is scored on the ordinate with red (RNA) fluorescence on the abscissa. Panel A shows data from the control non-stimulated cells (G_{1Q}). Stimulated cells that have incorporated BrdUrd are enclosed in the dashed polygon in panel B. Cells in transit to G1 (G_{1T}) have G1 RNA levels but have not entered S phase and have not incorporated BrdUrd, hence their green fluorescence is not quenched. Redrawn from Darzynkiewicz et al. (1978).

the division cycle had incorporated the analogue. These cells are included in the dashed polygon, and the G1A and G1B compartments clearly have reduced green (DNA) fluorescence compared with non-stimulated cells, labelled G1Q. The G1T cells, in transit from G_0 to G1A, are those with non-suppressed green fluorescence but with RNA levels comparable with those in G1A. These cells are stimulated as their RNA levels have increased above the G_0 cells, but they had not passed through DNA synthesis as their green fluorescence was not reduced. Studies by Traganos et al. (1979) with DMSO-induced senescence in Friend leukaemia cells revealed non-cycling cells in S and G2 as well as in G1, and these various studies enabled Darzynkiewicz, Tragnos and Melamed (1980) to identify twelve different compartments in the cell cycle.

Multiparameter analysis of lipopolysaccharide (LPS)–stimulated B lymphocytes, using acridine orange to give measurements of RNA and DNA with both 90° and forward light scatter, has shown that the post-mitotic, pre-DNA synthesis phase (G1) can now be subdivided into four distinct subphases: G_0, G1′, G1A and G1B (Kenter et al. 1986). Figure 9.33 shows three data sets where DNA (520RF AREA) is plotted against RNA (630RF AREA) for B cells stimulated 5 days previously with LPS. The data in the top, middle and bottom panels (respectively) were from control and colcemid-arrested cells and from cells 4.5 hours after colcemid release; each panel shows the characteristic G1A, G1B, S-phase and G2+M pattern. Our initial interpretation of the data in Figure 9.33 was that with colcemid (middle panel) G1A emptied into G1B, the latter emptied into S phase which subsequently emptied into G2+M (where the population was blocked), and that there was a residual non-cycling G_0 population represented by the cells with lowest RNA content. This interpretation seemed to be substantiated by the data in the bottom panel of Figure 9.33, obtained 4.5 hours after colcemid was washed off. There was a relative increase in the proportion of G1A and G1B cells which were presumed to have arisen from a resumption of progression through the cell cycle, with division at mitosis refeeding the G1A compartment. This interpretation, however, was not entirely satisfactory as there did not seem to be progression through S and G2+M.

The data were then re-examined in the DNA (520RF AREA) versus 90° lightscatter (488RS WDTH) data space, which is shown in Figure 9.34. The left and right columns give perspective 3-D views and conventional contour plots, respectively, where the top, middle and bottom panels correspond to those in Figure 9.33. The control data (top panels) show two peaks in G1 distinguished by their high and low 90° light scatter, with small proportions in S and G2+M. The same overall pattern was seen after colcemid (middle panels) with two exceptions. First, there was the expected accumulation of cells with G2+M DNA content. Second, and unexpectedly, there was an increase in the proportion of cells with G1 DNA content which exhibit low 90° scatter signals. After release from colcemid (bottom panels) the proportion of cells in S and G2+M remained about the same, but there was a decrease in the G1 DNA content population with low 90° scatter. The three data sets were gated as shown in the right

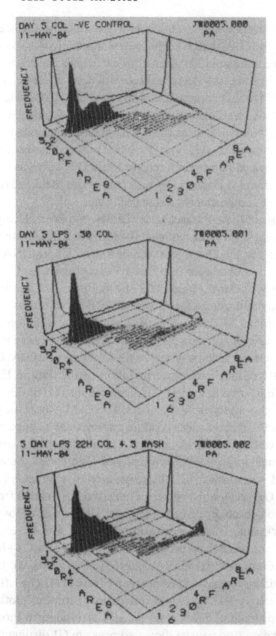

Figure 9.33. Acridine orange staining for RNA (630RF AREA) versus DNA (520RF AREA) in B cells stimulated 5 days previously with LPS. The top panel shows the control data; the middle and bottom panels (respectively) show the results of colcemid exposure and its withdrawal.

column of Figure 9.34 to give the proportions in G1' (region 1, defined as G1 DNA content with low 90° scatter), G1A plus G1B (region 2), G2+M (region 3) and S (region 4). The region-2 data were then analysed in the RNA × DNA data space to obtain the proportions in G1A and G1B; a summary of these data is

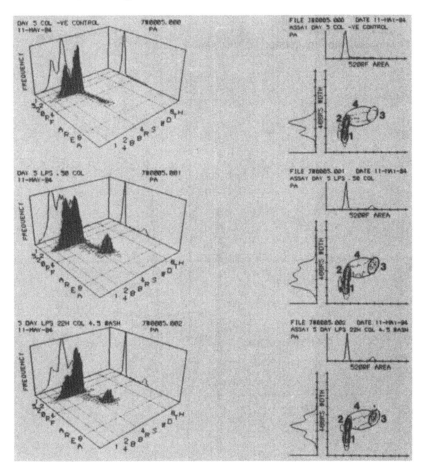

Figure 9.34. DNA (520RF AREA) versus 90° light scatter (488RS WDTH) for Figure 9.33 data. The right column shows conventional contour plots with four gated regions used for data analysis.

shown in Figure 9.35. With colcemid there was a relative increase of cells in G1′ to G1A and G1B.

Thus, the large spikes in the three panels of Figure 9.33 were not G_0 cells but rather G1′ cells plus some in G1A. Furthermore, the three ill-defined but distinct RNA distributions in the top panel of Figure 9.33 represent cells in G1′, G1A and G1B respectively. This was confirmed by two further pieces of evidence. First, unstimulated cells die and lyse with a half-time of about 18 hours, and after 5 days in culture there were no detectable unstimulated G_0 cells. Second, if the G1′ cells had been G_0 cells then the stimulated cells within the cell cycle would have had to double in 4.9 hours after release from colcemid in order to attain the proportions seen in the various phases 4.5 hours after colcemid was washed off. The doubling time of the population, from cell counts, was about 36 hours. Thus, this new G1′ compartment, which may be similar or

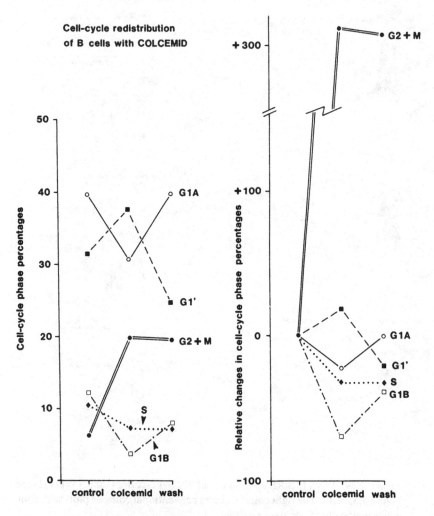

Figure 9.35. Summary of colcemid-induced changes in stimulated spleen B cells showing the unexpected transit block between G1′ and G1A and its rapid reversal following colcemid removal. The expected G2+M block is also apparent, but this was not reversed within the 4.5 hours after colcemid removal.

identical with the G1T compartment of Darzynkiewicz et al. (1980), is within the cell cycle and is blocked by colcemid.

Kenter and Watson (1987) also used the acridine orange technique in conjunction with Southern blotting (Southern 1975) to study the $S\mu$ to $C\mu$ immunoglobin heavy chain switching that occurs when stimulated spleen B cells convert from being IgM expressors to IgG secretors. The proportions of the population in G_0, G1′, G1A, G1B and S+G2+M were determined at 24-hour intervals for four days after stimulation with LPS and dextran sulphate. The model presented in Section 9.3 was extended to include these extra subdivisions in G1,

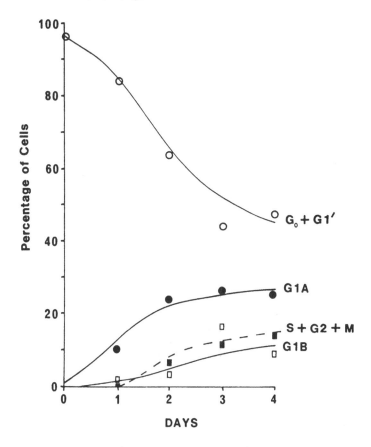

Figure 9.36. Computer model–predicted data (curves) versus experimentally determined proportions of B cells in G_0+G1' (o), G1A (•), G1B (□) and S+ G2+M (■) at the times (in days) shown on the abcissa after LPS and dextran sulphate stimulation.

and the various phase times with their standard deviations were estimated. Experimental data represented by points, and computer-predicted results represented by curves, are shown in Figure 9.36. The conversion from IgM expression to IgG secretion is accompanied by rearrangements within the immunoglobulin heavy chain locus, which are manifest by a loss of the $S\mu$ DNA signal from restriction enzyme digest patterns; the loss of this signal was analysed at intervals after stimulation by Southern blotting. Computer simulations were then carried out with the results from the parallel cell-cycle kinetic analysis to predict the proportion of $S\mu$ DNA signal remaining after stimulation. These results predicted that the $S\mu$ to $C\mu$ heavy chain switch would occur at approximately 85% into the cell cycle, placing it in mid S phase on the first pass through the cell cycle after stimulation. This is an interesting result, as it corresponds to the time in S phase at which the immunoglobulin locus is replicated; the results

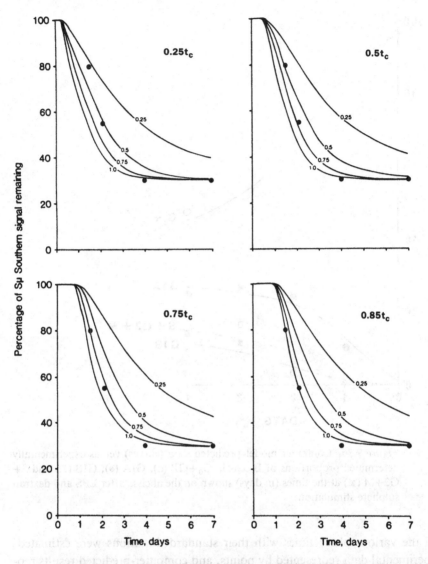

Figure 9.37. Computer-predicted DNA rearrangement of $S\mu$ signal (curves) versus experimental data (points) obtained from Southern analysis of $S\mu$ signal loss. The curves were simulated with switch probabilities of 0.25, 0.5, 0.75 and 1.0 at times corresponding to early G1 ($0.25t_c$), mid G1 ($0.5t_c$), G1/S interface ($0.75t_c$) and mid S phase ($0.85t_c$). The best correspondence between experimental and simulated data occurred at $0.85t_c$ with unit switch probability.

are shown in Figure 9.37. The predictions were confirmed experimentally by treating stimulated cells with aphidocholine, which blocks cells in early S phase. This also blocked the loss of the $S\mu$ signal, but the switch was manifest by loss of the signal some 6 hours after release from aphidocholine.

10

Dynamic cellular events

Measurement of dynamic cellular events in intact cells under near physiological conditions, irrespective of the assay system, is arguably one of the most powerful techniques in cell biology, as physiological and pathological processes are not static but are continuously variable. It is becoming increasingly obvious that many of the initial events involved in cell proliferation, "activation" of all types and differentiation control take place within a time frame of seconds, but the ultimate effects of those initiating events (e.g. growth-factor stimulation, membrane calcium flux, pH changes and carcinogen interaction) may only become apparent some days, weeks or even years later. Clearly, an understanding of the dynamics of inappropriate or abnormal initiating events will be of crucial importance in attempts to modulate pathological states. This chapter addresses flow cytometric analysis procedures to measure dynamic events.

10.1 Incorporation of time

Incorporation of time in the data base is the most important single feature in the measurement of dynamic cellular processes using flow cytometry, and a number of different techniques have been implemented for this. Watson et al. (1977) used a continuous interrupted sampling technique, where the cells were continuously flowing through the instrument. The data-acquisition computer was instructed to record for 5 seconds and then wait for 10 seconds before re-commencing acquisition. During the interval between recordings the median of the population was calculated and printed out. This method had resolution problems, particularly for fast reactions, as the flourescence from the population was increasing during the data acquisition interval and hence the medians tended to be overestimated. Martin and Swartzendruber (1980) overcame this very elegantly by incorporating time directly into the data base. Their initial method used a voltage ramp generator, where the potential increased linearly with time and a recording of the voltage was made from the ramp generator each time a cell was analysed in the flow chamber. This method had one disadvantage, namely that time could be recorded only over an interval of about

15 minutes. However, immediately following the introduction of continuous time recording by Martin and Swartzendruber (1980) we incorporated a time stamp from the data-acquisition computer directly into the data base. Hence, events were timed automatically with 50-ms resolution by the computer clock and this time, in the form of the number of clock "ticks" (one tick per 50 ms), was recorded for each event analysed.

Keij et al. (1989) have incorporated "pseudo" time into list-mode data by inference from the computer-recorded duration of the run. The latter was divided sequentially into 255 equal channels, producing a time-related factor that was proportional to time. This is not the ideal method for incorporating time into the data base, as extreme care must be taken to keep the flow rate constant. However, it is the only compromise possible for commercial instruments that do not include the facility to incorporate time in the data base from the computer clock.

10.2 Classical enzyme kinetics

A general schematic for the conversion of substrate to product by enzymatic action is shown in Figure 10.1. Substrate combines with enzyme to form an enzyme–substrate complex with rate constant k_1. This may revert to free enzyme and substrate with rate constant k_{-1} or dissociate into product and free enzyme with rate constant k_2. The differential equations describing the reaction are as follows:

$$\frac{\delta[\text{ES}]}{\delta t} = k_1[\text{E}][\text{S}] - k_{-1}[\text{ES}] - k_2[\text{ES}], \tag{10.1}$$

$$\frac{\delta[\text{P}]}{\delta t} = k_2[\text{ES}], \tag{10.2}$$

$$\frac{\delta[\text{E}]}{\delta t} = -k_1[\text{E}][\text{S}] + k_2[\text{ES}], \tag{10.3}$$

$$\frac{\delta[\text{S}]}{\delta t} = -k_1[\text{S}], \tag{10.4}$$

$$[\text{E}_0] = [\text{E}] + [\text{ES}]. \tag{10.5}$$

If you think you don't understand these equations it's only because you are not familiar with the symbols, so if you know all about these things then go on to the next section. But if you don't then we'll go through the first equation, (10.1), in detail, starting on the left. $\delta[\text{ES}]/\delta t$ means the rate of change of ES, the enzyme–substrate complex. In this context it is the "speed" at which ES is being formed. Rates of change are less daunting if they are put into the context of everyday experience. The road sign shown in Figure 10.2 means that there is

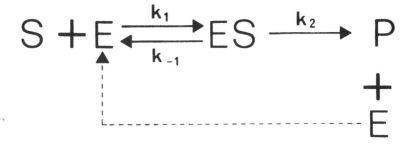

Figure 10.1. Schematic for conversion of substrate to product via an intermediate enzyme–substrate complex.

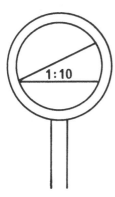

Figure 10.2. Road sign for 1-in-10 gradient.

a gradient of 1 in 10. Thus, for every 10 meters you travel horizontally you also go up or down (depending on your direction) by 1 meter. This sign could have been written $\delta\mathrm{Road}/\delta D = 0.1$ where D is distance. Clearly, if you are moving then distance and time are related, so the D in the expression δD could be replaced by t (for time) to give $\delta\mathrm{Road}/\delta t = ?$. I've replaced the 0.1 of $\delta\mathrm{Road}/\delta D = 0.1$ by ? because the time it will take you to get up or down the incline in the road is dependent on your speed. $\delta\mathrm{Road}/\delta D$ is a constant of the road at that particular point, but $\delta\mathrm{Road}/\delta t$ depends on you. If you are a jogger and are running up the incline then your speed, $\delta\mathrm{Road}/\delta t$, depends on the incline – including both the constant of the incline, $\delta\mathrm{Road}/\delta D$, and its length – as well as your fitness and desire to get up the incline.

Returning to equation (10.1), we can see that the speed at which ES is being formed is given by the first term on the right-hand side, $k_1[\mathrm{E}][\mathrm{S}]$, where $[\mathrm{E}]$ and $[\mathrm{S}]$ respectively are the concentrations of free enzyme and substrate which are available to react, and where k_1 is the rate constant for that reaction (which is analogous with $\delta\mathrm{Road}/\delta D$, the incline constant of the road). The speed at which ES is dissociating is determined by two factors. The second term $k_{-1}[ES]$ is the speed at which the ES complex is reverting to free enzyme E and substrate S with rate constant k_{-1}, and the term has a minus sign in front of it to indicate that ES is dissociating. The last term $k_2[ES]$ is the speed at which ES is dissociating into product and free enzyme with rate constant k_2. This term is also preceded by a minus sign to represent dissociation of the complex ES. The second rate equation (10.2), $\delta[P]/\delta t = k_2[ES]$, represents the speed at which the product P is being formed from ES; this occurs at the same rate at which ES is dissociating into product and free enzyme, the last term in the first rate equation, but the sign is now positive. The concentration of free enzyme at any given time is obtained by rearranging equation (10.5) to give $[E] = [E_0] - [ES]$, where $[E_0]$ is the initial enzyme concentration and where ES represents enzyme bound to substrate. We can now substitute for [E] in equation (10.4) to give

$$\frac{\delta[ES]}{\delta t} = k_1([E_0] - [ES])[S] - k_{-1}[ES] - k_2[ES],$$

which upon rearrangement yields

$$\frac{\delta[ES]}{\delta t} = k_1[E_0][S] - k_1[ES][S] - [ES](k_{-1} + k_2)$$
$$= k_1[E_0][S] - [ES](k_1[S] + k_{-1} + k_2). \tag{10.6}$$

When a steady state exists, the speed at which ES is being formed is equal to the speed at which it is dissociating; thus rate of change of [ES], $\delta[ES]/\delta t$, will be zero and so equation (10.6) reduces to

$$[ES](k_1[S] + k_{-1} + k_2) = k_1[E_0][S],$$

from which we obtain the following expression for [ES] in terms of the remaining parameters:

$$[ES] = \frac{k_1[E_0][S]}{k_1[S] + k_{-1} + k_2}.$$

We now divide through the right-hand side by k_1, which gives

$$[ES] = \frac{[E_0][S]}{[S] + \left[\dfrac{k_{-1} + k_2}{k_1}\right]}.$$

The expression $(k_{-1} + k_1)/k_1$ is the Michaelis constant K_m, so the above equation reduces to

$$[ES] = \frac{[E_0][S]}{[S] + K_m}. \tag{10.7}$$

We now return to equation (10.2) which when rearranged gives

$$[ES] = \frac{\delta[P]}{\delta t}\frac{1}{k_2}. \tag{10.8}$$

Equations (10.7) and (10.8) can now be equated, which eliminates [ES] to give the rate of change of [P]:

$$\frac{\delta[P]}{\delta t} = \frac{k_2[E_0][S]}{[S]+K_m}.$$ (10.9)

$\delta[P]/\delta t$ is the rate at which product is formed and represents the reaction velocity v. At infinite substrate concentration at the very start of the reaction (this is theoretical), when all the enzyme is in its free form $[E_0]$, the reaction velocity will be at its maximum V, which is given by $k_2[E_0]$; the equation may be written as

$$v = V\frac{[S]}{[S]+K_m}.$$ (10.10)

This gives the initial reaction velocity v for a given substrate concentration [S] with the Michaelis constant K_m for that particular reaction.

Plots of initial reaction velocity v versus increasing substrate concentration are shown in Figure 10.3, the so-called Michaelis–Menten rectangular hyperbola, although Michaelis and Menten (1913) did not describe this mathematical analysis. As the substrate concentration increases, the reaction velocity becomes closer and closer to the theoretical maximum V, and K_m is the substrate concentration where $v = V/2$. The initial velocities are obtained by regression analysis of the initial slopes of enzyme progress curves obtained with increasing substrate concentrations; a representation is shown in Figure 10.4. It is difficult to estimate the Michaelis constant directly from the type of plot shown in Figure 10.3, as V cannot be defined with any degree of precision because it is impossible to perform an experiment with infinite substrate concentration. Moreover, there will always be some uncertainty in the results, which makes an estimate from the type of plot in Figure 10.3 even more difficult. However, help is at

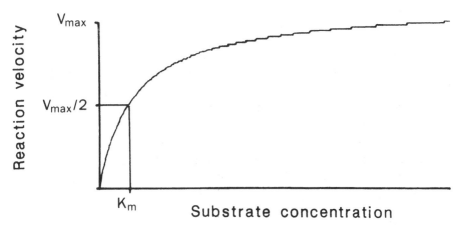

Figure 10.3. The Michaelis–Menten rectangular hyperbola of reaction velocity versus substrate concentration.

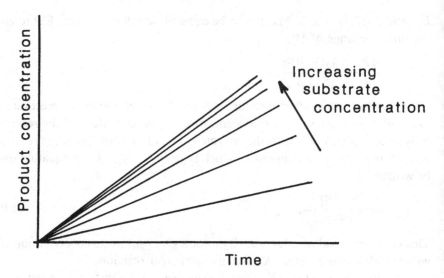

Figure 10.4. Enzyme progress curves, where product concentration is plotted against time. As the substrate concentration increases so the slopes approach the limit of V_{max}.

hand with a variety of linear transforms of the Michaelis–Menten rectangular hyperbola.

The first of these is the Lineweaver–Burke double reciprocal $1/v$-versus-$1/[S]$ plot (Lineweaver and Burke 1934). If we take the reciprocal of the Michaelis–Menten hyperbola (10.10) we obtain

$$\frac{1}{v} = \frac{[S] + K_m}{V[S]}.$$

We now rearrange both components of the right-hand side to give

$$\frac{1}{v} = \frac{[S]}{V[S]} + \frac{K_m}{V[S]} = \left(\left[\frac{K_m}{V} \right] \times \frac{1}{[S]} \right) + \frac{1}{V}.$$

This reciprocal equation is now of the form $y = mx + c$, where $1/v$ is equivalent to y, K_m/V is equivalent to the slope m, and $1/V$ is the intercept c on the Y-axis. Thus, by plotting $1/v$ versus $1/[S]$ as in Figure 10.5 and carrying out a regression analysis, we obtain a straight line which cuts the Y-axis at $1/V$; hence, $V = 1/c$, the reciprocal of the intercept. As the slope m equals K_m/V we have $K_m = m/c$, and the kinetic parameters of the reaction are fully defined.

The second linear transform is due to Eadie (1942) and Hofstee (1952). The hyperbola (10.10) is multiplied through on both sides by $[S] + K_m$ to give

$$v([S] + K_m) = V[S],$$

$$v[S] + vK_m = V[S].$$

We now divide through by $[S]$ and rearrange, which yields

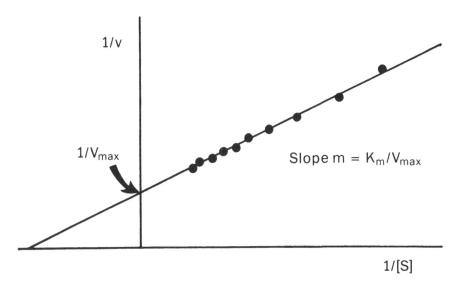

Figure 10.5. Lineweaver-Burke plot of $1/v$ versus $1/[S]$.

$$v = -K_m \frac{v}{[S]} + V.$$

This equation also is now in the form $y = mx + c$, where v is equivalent to y, the slope m is $-K_m$, x is equivalent to $v/[S]$, and the intercept c is equivalent to V. Thus by plotting v versus $v/[S]$ and performing regression analysis as shown in Figure 10.6, we obtain K_m and V.

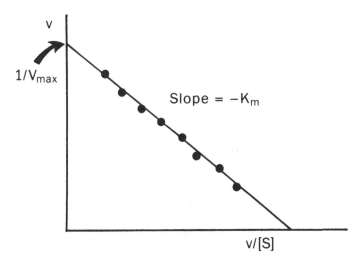

Figure 10.6. Eadie–Hofstee plot of v versus $v/[S]$.

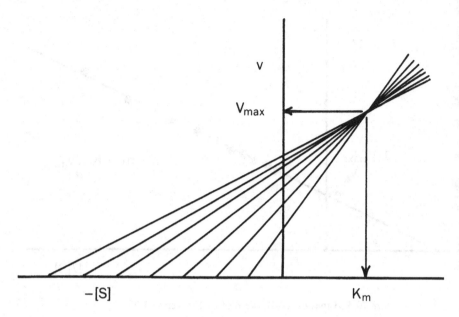

Figure 10.7. Direct linear plot of Eisenthal and Cornish-Bowden.

The third transform is due to Eisenthal and Cornish-Bowden (1974), who rearranged the hyperbola to give the maximum velocity V in terms of the remaining parameters. This is effected by rearranging the Eadie–Hofstee transform to give

$$V = v + (v \times K_m)/[S].$$

For each value of v associated with its substrate concentration $[S]$, a line can be plotted which crosses the abscissa at $-[S]$ and the ordinate at v. With a perfect data set the various lines for each v and $[S]$ combination will all intersect at a point with (x, y)-coordinates of (K_m, V). This is the direct linear plot of Eisenthal and Cornish-Bowden, which is illustrated in Figure 10.7.

10.3   Flow cytoenzymology

Flow cytoenzymology is now an accepted term for studying enzyme reaction kinetics in intact cells using flow systems. This term should be used when describing results obtained with these instruments, as such results can differ considerably from those obtained with classical techniques. The first non-kinetic description of enzyme measurements by flow cytometry (Hulett et al. 1969) exploited the phenomenon of flourochromasia (Rotman and Papermaster 1966), in which the lipophilic, membrane-permeable fluorogenic substrate fluorescein diacetate (FDA) was converted to the comparatively polar fluorescent product fluorescein by intracellular esterases. One of the first kinetic assays

INTRACELLULAR   ENZYMES

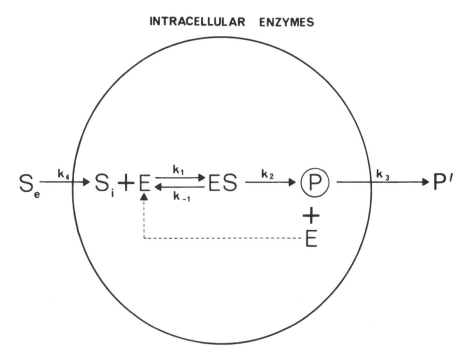

Figure 10.8. Schematic for intracellular conversion of substrate to product. Only product that is cell-associated is "seen" by flow cytometers.

where the development of fluorescence with time was recorded also used FDA (Watson et al. 1977). Since then there have been considerable advances which will continue with the further developments of probes.

10.3.1 Cytoplasmic enzymes

A schematic for the conversion of substrate to product by enzyme re-action within whole cells is shown in Figure 10.8. This differs from the schematic shown in Figure 10.1 in that substrate external to the cell, S_e, must first cross the external membrane to enter an intracellular pool S_i and then interact with the enzyme generating the reaction product, which can then be measured if it remains inside or associated with the cell. Unlike classical biochemical as-says, only product that is cell-associated, P, is "seen" by the instrument. Any product that diffuses away from the cell, P', is not scored. The ideal fluoro-genic substrate for flow enzymology is lipophilic, non-toxic and non-fluorescent. During the reaction this is converted to a highly polar fluorescent product. Sub-strate lipophilicity generally is required for rapid transport across the external cell membrane and high polarity of the product is desirable in order to "trap" the product within the cell, but few substrates meet all these criteria. However, there is now a considerable variety of such substrates based on a number of dif-ferent fluorophores including fluorescein, rhodamine, naphthol, naphthylamine, quinolines, methylumbelliferone and monochlorobimane. These have been used for studying esterases, lipases, sulphatases, phosphatases, glucuronidases, transferases, peroxidases, peptidases, transpeptidases, galactosidases, arylamidases, glucosidases and glutathione S-transferases, and the list is growing all the time.

In general, the fluorescein- and rhodamine-based substrates and monochlorobimane are either weakly or non-fluorescent, which is ideal. In contrast, the 4-methylumbelliferone conjugates are generally fluorescent, but both their excitation and emission wavelengths are much shorter than those of the released product 4-methylumbelliferone; with correct optical design and/or electronic compensation any "overlap" can be minimised or eliminated. Techniques using the naphthol derivatives involve trapping the released product within the cell by coupling with 5-nitrosalicyl-aldehyde, which forms an insoluble fluorescent complex with an emission spectrum that is shifted into the red (Dolbeare and Smith 1971). A comprehensive list of substrates and the reactions to release the associated fluorophores is not in order here, and the reader is referred to the Molecular Probes Catalogue (Molecular Probes Inc., Eugene, Oregon, USA) and an excellent volume, *Applications of Fluorescence in the Biomedical Sciences* (Lansing-Taylor et al. 1986).

Esterases

The continuous interrupted sampling technique (Watson et al. 1977) with FDA as substrate for esterases was used to show that cells in late plateau–phase EMT cultures, in which more than 95% of the population was arrested in a $G1/G_0$ state, exhibited considerably higher esterase activity than exponentially growing cells (Watson, Workman and Chambers 1978). The progress curves are shown in Figure 10.9, with data from the plateau phase and exponentially growing population in panels A and B, respectively. The micromolar concentrations of substrate are shown against each progress curve. Figure 10.10 shows the substrate-dependent velocity plots associated with these data, which theoretically should follow the Michaelis–Menten rectangular hyperbola (Michaelis and Menten 1913). There was a highly abnormal double-sigmoid pattern for the plateau-phase cells, but the exponentially growing cells exhibit less abnormal kinetic behaviour. Krisch (1971) has reported abnormal kinetic behaviour in certain esterase preparations, and it has been suggested that the reaction mechanism may need two or more interacting catalytic sites. Substrate activation, where the reaction proceeds most rapidly when more than one substrate molecule is bound to a single enzyme molecule, may also be involved (Adler and Kistiakowsky 1961). Furthermore, it is highly probable that more than one esterase is involved with the hydrolysis of FDA. Guibault and Kramer (1966) have shown that FDA hydrolysis is catalysed by a number of enzymes including α- and γ-chymotrypsin and lipase. However, additional factors must also be considered for the abnormal kinetic behaviour shown in Figure 10.10. First, cellular and subcellular permeability barriers are likely to exist in the intact cell, which could

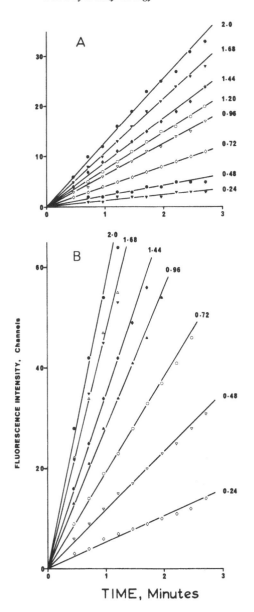

TIME, Minutes

Figure 10.9. Enzyme progress curves of esterase activity in log- and plateau-phase EMT6 cultures using FDA as substrate (panels A and B, respectively). The micromolar concentrations of substrate are shown against each progress curve.

limit the availability of substrate at the enzyme site. Second, there could be active-transport mechanisms for the substrate with intracellular accumulation. These two factors each would result in a difference between the substrate concentration external to the cell and at the site of enzyme reaction, and all of

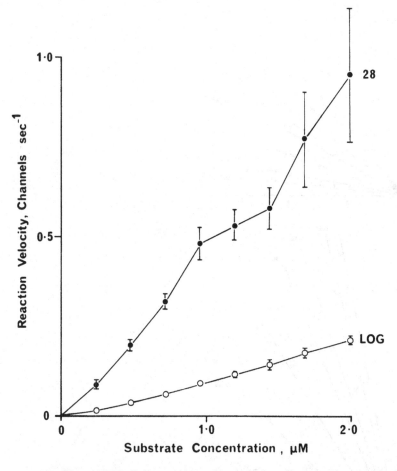

Figure 10.10. Substrate-dependent velocity plots associated with the data in Figure 10.9, which should theoretically follow the Michaelis–Menten rectangular hyperbola.

these various factors could contribute to the upward concavity ("lag kinetics") seen for plateau-phase cells in Figure 10.10.

Glutathione S-transferase(s)

Glutathione (GSH) is a critical determinant in control of cellular response to anticancer drugs and radiation. Depletion of protective GSH results in sensitisation. Kosower et al. (1979) were the first to use fluorescent bimanes to study intracellular thiols. Rice et al. (1986), using flow cytometry, have shown that monochlorobimane (mClB) can act as a reporter molecule in GSH metabolism. As an extension to this, we developed a sensitive multiparametric flow cytoenzymological assay using continuous time measurements to determine reaction kinetics for conjugation of mClB with GSH in populations of intact viable

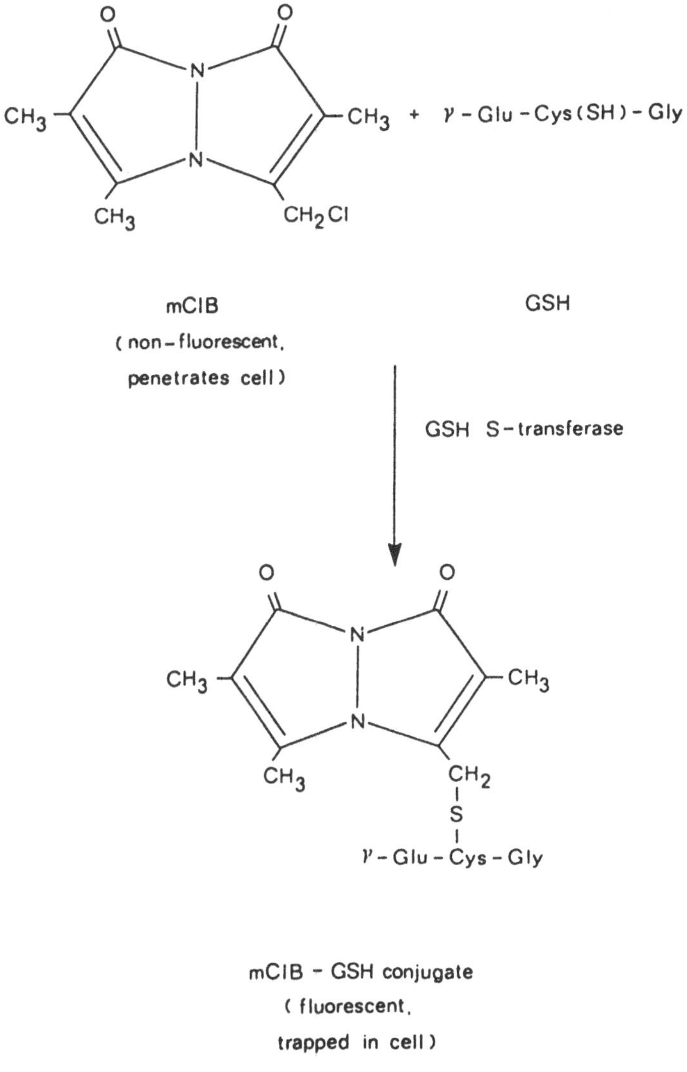

Figure 10.11. Reaction of monochlorobimane with the tripeptide glutathione, mediated by transferases to produce the insoluble fluorescent conjugate.

cells catalysed by the enzyme family, the glutathione S-transferases (Workman and Watson 1987; Workman, Cox and Watson 1988). Monochlorobimane is a non-fluorescent probe which after conjugation with GSH results in a relatively insoluble fluorescent complex excited by UV light with emission in the blue (460–510 nm). The reaction is illustrated in Figure 10.11, and the reaction rate is dependent on mClB concentration, intracellular GSH content and the activity of the transferase(s). A cartoon of the reaction, which includes the rate constants and the possibility for GSH production during the reaction, is shown

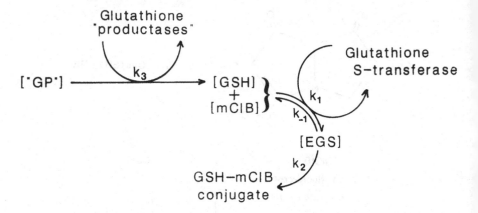

Figure 10.12. Schematic for the conversion of mClB and GSH to the conjugate by transferases. An additional GSH production pathway (GP) is included for completeness.

in Figure 10.12. The differential rate equations describing the system are:

$$\frac{\delta[\text{EGS}]}{\delta t} = k_1[\text{E}][\text{G}][\text{S}] - k_{-1}[\text{EGS}] - k_2[\text{EGS}],$$

$$\frac{\delta[\text{P}]}{\delta t} = \qquad\qquad\qquad + k_2[\text{EGS}],$$

$$\frac{\delta[\text{S}]}{\delta t} = -k_1[\text{E}][\text{G}][\text{S}] + k_{-1}[\text{EGS}],$$

$$\frac{\delta[\text{E}]}{\delta t} = -k_1[\text{E}][\text{G}][\text{S}] \qquad\qquad + k_2[\text{EGS}],$$

$$\frac{\delta[\text{G}]}{\delta t} = -k_1[\text{E}][\text{G}][\text{S}] + k_{-1}[\text{EGS}] \qquad\qquad + k_3[\text{GP}],$$

$$[\text{E}] = [\text{E}_0] - [\text{EGS}].$$

The shorthand notation in the equation matrix is as follows. [E], [G] and [S] represent the concentrations of free enzyme, glutathione and substrate, respectively. [EGS] represents the concentration of the complex of enzyme plus glutathione plus substrate. [E] is the concentration of free enzyme at any given time and $[\text{E}_0]$ is the initial concentration of enzyme. [GP] represents the glutathione production system, which is considerably more complex than represented here, but this very simplified version has been included for completeness.

This system is more complicated than that shown in Section 10.2 as $\delta[\text{EGS}]/\delta t$ is dependent on three concentrations, namely [E], [S] and [G]. Hence, this is a third-order system and is not readily amenable to the type of solution presented for the second-order system in Section 10.2. However, this is no problem

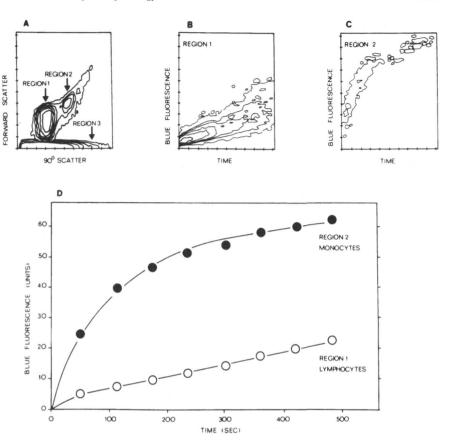

Figure 10.13. Ficoll–Paque lymphocyte-enriched preparation obtained from heparinised whole blood of a normal subject.

in this day and age, as there are numerical integrating algorithms available to tackle exactly this type of differential equation matrix.

The progress curves generated from this equation matrix reach an asymptote when all intracellular GSH is converted. Figure 10.13 shows the results from a Ficoll–Paque lymphocyte-enriched preparation obtained from heparinised whole blood of a normal subject, which was mixed with 120 μM mClB plus propidium iodide at a concentration of 50 μg ml^{-1} (Workman et al. 1988). Forward and 90° scatter signals were collected together with blue and red fluorescence. The latter is also excited by UV from the DNA of dead cells, which do not exclude propidium iodide. Panel A shows forward versus 90° scatter, where three regions can be defined. Regions 1 and 2 correspond to lymphocytes and monocytes respectively. Region 3 represents debris and propidium iodide–positive dead cells. Panels B and C show the accumulation of blue fluorescence with time as frequency contour plots, where it can be seen that monocytes (panel C) exhibited greater activity than lymphocytes (panel B), which is consistent with larger quantities of GSH and/or GSH S-transferase. A more conventional display, derived from the data in panels B and C, shows medians of the distributions at discrete time intervals plotted against time in panel D. The dead cells and debris in region 3 exhibited no activity. The data in Figure 10.13 represent a 5-dimensional set, as two scatter and two fluorescence measurements were recorded with the fifth parameter, time.

We have consistently found that monocytes reach a fairly well-defined asymptote but lymphocytes and some tumour cells frequently do not. One explanation is that monocytes may not have the capacity to replenish intracellular GSH levels and as a result their progress curves reach a well-defined asymptote. However, any cells having the capacity to generate glutathione in response to depletion during the reaction will not reach a well-defined asymptote. If this is true then the initial phase of the reaction may reflect GSH transferase activity more accurately and in the later phases the continuing increase in blue fluorescence levels would reflect GSH production. A second explanation is substrate non-specificity. Monochlorobimane plus any thiol can act as substrates for the appropriate transferases. Hence, we might expect departures from the predicted progress curve in cells that are thiol rich, other than in GSH, including antibody-producing cells.

10.3.2 Membrane enzymes

An automated fluorimetric method for assaying alkaline phosphatase using 3-O methyl fluorescein phosphate was developed by Hill, Sumner and Waters (1968). An adaptation of this technique has been applied to intact cells using flow cytometry by Watson, Workman and Chambers (1979). Again, the continuous interrupted method was used. The results for EMT6 cells are shown in Figure 10.14. These are obviously very different from those for intracellular esterases shown in Figure 10.9, and they presented a considerable interpretative problem which was not resolved until a sample was viewed under the fluorescence microscope. It was then seen that the fluorescence was concentrated as "halos" at the external cell membrane, with no fluorescence from the interior of the cells. Subsequent incubation of cells with product, 3-O methyl fluorescein, failed to demonstrate entry into intact cells. The viability of EMT6 cells decreases considerably after three hours of incubation with protein-free phosphate-buffered saline (PBS). In a sample of cells so treated it was found that the fluorescence from 3-O methyl fluorescein was no longer located at the cell surface but was being emitted from granular structures surrounding the nucleus. This fluorescence was considerably more intense than that emitted from the cell surface, so much so that the latter could no longer be seen. These various data suggested that substrate hydrolysis in intact viable cells takes place at or within the cell by phosphatases located in the plasma membrane, and that the product was lost from the immediate vicinity of the cells by diffusion and consequently

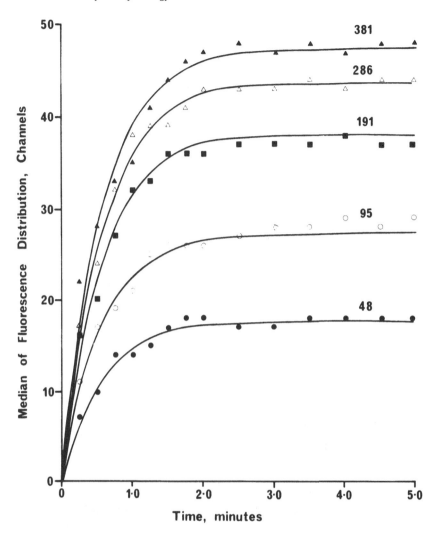

Figure 10.14. Progress curves for hydrolysis of 3-O methyl fluorescein phosphate. The μM concentrations of substrate are shown adjacent to each curve.

was not detected by the instrument. Thus a steady state would be reached in which the rate of production of fluorescent product would equal the rate of loss, giving rise to asymptotic fluorescence responses from the population at each substrate concentration.

Assuming that this hypothesis is correct, it is possible to modify equation (10.9) describing the rate of change of product (see Section 10.2) by adding the term $-k_3[P]$ to the right-hand side, which gives

$$\frac{\delta[P]}{\delta t} = \frac{k_2[E_0][S]}{[S] + K_m} - k_3[P],$$

where k_3 is the rate constant for product loss assuming first-order kinetic processes. Rearrangement of this equation gives

$$\frac{\delta[P]}{\delta t} + k_3[P] = \frac{k_2[E_0][S]}{[S] + K_m},$$

and on integration we obtain the following inverted exponential (the full derivation of which is given in Appendix 9):

$$F(t) = F_\infty \times \frac{[S]}{[S] + K_m} \times (1.0 - \exp(-k_3 \times t)). \tag{10.11}$$

$F(t)$ is the fluorescence response at time t, F_∞ is the theoretical maximum asymptotic response at infinite substrate concentration, K_m is the Michaelis constant and k_3 is the rate constant for product loss assuming a first-order kinetic process. When a steady state exists (i.e. beyond 2.5–3 min in Figure 10.14), the term $\exp(-k_3 \times t)$ in equation (10.11) will tend to zero and the asymptotic fluorescence response will be given by the expression $F_\infty \times [S]/([S] + K_m)$ and will vary with substrate concentration. F_∞ can be eliminated from this expression by taking ratios; thus

$$\frac{F_2(t_\infty)}{F_1(t_\infty)} = \frac{[S_2]}{[S_1]} \times \frac{[S_1] + K_m}{[S_2] + K_m} = R_{12}, \tag{10.12}$$

where $F_1(t_\infty)$ and $F_2(t_\infty)$ are the asymptotic fluorescence values associated with substrate concentrations $[S_1]$ and $[S_2]$ (respectively) and where R_{12} is the ratio of $F_2(t_\infty)$ to $F_1(t_\infty)$. Equation (10.12) can be rearranged to give:

$$R_{ij} = K_m \times \left[\frac{1}{[S_i]} - \frac{R_{ij}}{[S_j]} \right] + 1,$$

where for N substrate concentrations i varies from 1 to $(n-1)$ and j varies from $(i+1)$ to N, to give a triangular matrix for the ratios containing $N(N-1)/2$ values. Thus, by plotting R_{ij} against $(1/[S_i]) - (R_{ij}/[S_j])$, a line with slope K_m is obtained which intersects the ordinate at unity; see Figure 10.15. In three separate analyses, values of 110.0 ± 31 μM, 122.3 ± 31 μM and 120.5 ± 24 μM were obtained, where the limits were calculated at two standard errors and in all cases the ordinate intercept did not differ from unity, $p > 0.1$. It can be seen from equation (10.12) that if F_∞ is the asymptotic fluorescence response at infinite substrate concentration $[S_\infty]$ then

$$\frac{F_\infty}{F_n} = \frac{[S_\infty]}{[S_n]} \times \frac{[S_n] + K_m}{[S_\infty] + K_m},$$

where F_n is the asymptotic response at $[S_n]$. The term $[S_\infty]/([S_\infty] + K_m)$ is unity, so $F_n \times ([S_n] + K_m) = F_\infty \times [S_n]$. By plotting $F_n \times ([S_n] + K_m)$ against $[S_n]$ a line of slope F_∞ is obtained which intersects the origin; see Figure 10.16. Thus F_∞ can be obtained after a value for K_m is found. The maximum reaction velocity

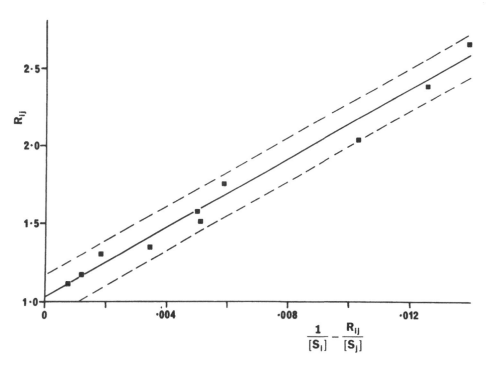

Figure 10.15. Derivative plots of R_{ij} versus $(1/[S_i]) - (R_{ij}/[S_j])$; see text. The slope of the regression lines gives the value of K_m, and the 95% confidence limits are shown by the dashed lines. The ordinate intercept does not differ from unity, $p > 0.1$.

V is equal to the product of $F_\infty \times k_3$, which can be be calculated after k_3 is defined. Equation (10.11) is an inverted exponential, and since K_m and F_∞ have been obtained we can use the linear transform described in Section 5.2.1 to obtain values for k_3 at each substrate concentration S. The transform for this particular problem is

$$\log_e\left[1 - \left(\frac{F(t)}{F_\infty} \times \frac{[S] + K_m}{[S]}\right)\right] = -k_3 \times t.$$

These regressions are shown in Figure 10.17. As we can see, the slopes k_3 are very similar and none of the intercepts differed from zero, $p > 0.999$ in all cases. As the values obtained for k_3 were all so similar and independent of [S] (which they should be), we now perform a simultaneous regression analysis to obtain a common value for k_3. This, and a subsequent common curve-fitting analysis, gave a rate constant of 1.98 min^{-1}, thus V can be defined in the arbitrary units of channels min^{-1}. The three analyses gave maximum reaction velocities of 107.0, 110.0 and 109.2 channels per minute. The curves shown in Figure 10.14 were calculated with the various parameters obtained here.

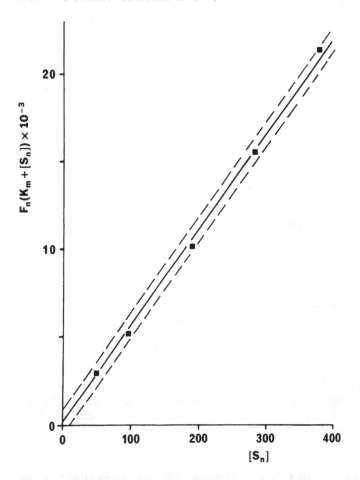

Figure 10.16. Regression of $F_n(K_m + [S_n])$ on $[S_n]$; see text. The slope gives F_∞, defined as the asymptotic fluorescence intensity at infinite substrate concentration. The 95% confidence limits are shown as the dashed lines, and the ordinate intercept does not differ from zero, $p > 0.1$.

The progress curves in Figures 10.9 and 10.14 are very different, and the major differences between the fluorogenic substrates and their products should be considered. First, 3-O methyl fluorescein phosphate is highly polar, ionised at physiological pH and hydrophilic. In contrast fluorescein diacetate is essentially non-polar, is not ionised at physiological pH and is lipophilic. Thus fluorescein diacetate will penetrate the cell without prior hydrolysis by any membrane esterases, whereas 3-O methyl fluorescein phosphate would not be expected to traverse the cell membrane unless there was either an active transport mechanism for it or it was dephosphorylated at the membrane. Second, 3-O methyl fluorescein, the reaction product of 3-O methyl fluorescein phosphate, is considerably less polar than fluorescein, the reaction product of FDA. Thus, if 3-O methyl fluorescein enters the cell we would expect the rate constant for leakage

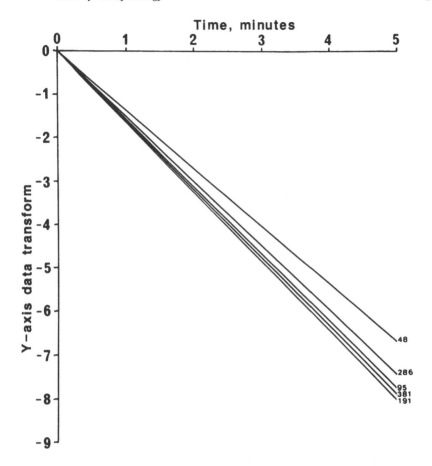

Figure 10.17. Regression lines for the individual inverted-exponential linear transform analyses of the data points shown in Figure 10.14.

from the cell to be greater than that for fluorescein. Studies with EMT6 cells have shown that fluorescein leaks out of these cells with a half-time of 7–8 min (Watson et al. 1979; Watson 1980), thus the leakage rate constant for 3-O methyl fluorescein is about 20 times greater than that for fluorescein. The direct observations made with the fluorescence microscope suggest that 3-O methyl fluorescein either does not enter the cell or diffuses out of the cell very rapidly (although some remains associated with the external membrane). Both possibilities are compatible with the magnitude of the loss rate constant, whatever the mechanism.

10.3.3 Inhibition kinetics

Chloroethylnitrosoureas (Cnus) are an important class of anti-tumour agent. Their exact mode of action is not known but chloroethylation of DNA and subsequent cross-linking has been implicated (Gibson, Mattes and Hartley 1985). However, protein carbamoylation may also be involved as organic

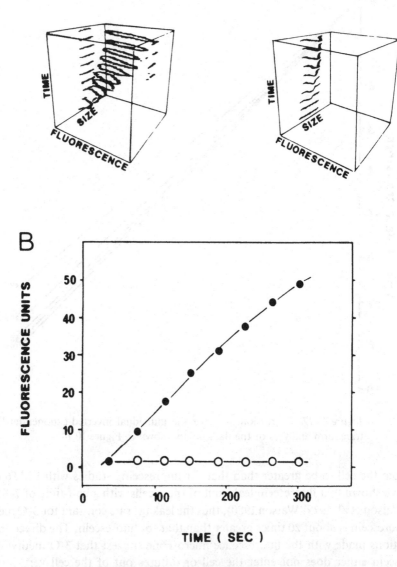

Figure 10.18. Inhibition of esterase activity by BCNU-derived isocyanate. The top two diagrams (panel A) show size versus fluorescence on the *X* and *Z* horizontal axes, with time (which increases upwards on these displays) on the *Y*-axis. Contour plots of size versus fluorescence for 10 equal "time-slices" are shown within these 3-dimensional data spaces. Top left is the control and top right shows the inhibition due to BCNU. Panel B shows the medians of the fluorescence distributions of each time-slice plotted against time.

isocyanates as well as alkylating species are formed when Cnus decompose in aqueous environments. Carbamoylating agents have been clearly shown to inactivate a number of enzymes, and inactivation of glutathione reductase has been used to determine the protein carbamoylation potential of Cnus (Babson and Reed 1978). Because of the susceptibility of serine hydrolases to isocyanate inactivation (Brown and Wold 1973), Paul Workman of our laboratories proposed that flow cytometric measurement of esterase activity might form a basis for the determination of intracellular carbamoylation by Cnu-derived isocyanates (Dive 1988; Dive, Workman and Watson 1987, 1988). This proved to be the case, and an example of esterase inhibition induced by BCNU-derived isocyanate is shown in Figure 10.18 (see figure caption for full explanation). By using a number of different drug doses as in Figure 10.19 it is possible to obtain an estimate of the drug concentration which causes a 50% reduction in FDA hydrolysis activity, defined as the I_{50} value. Using the I_{50} value it is possible to compare directly the inhibitory effects of a number of Cnus and related cytotoxic agents; such comparisons are discussed in Section 10.5.3.

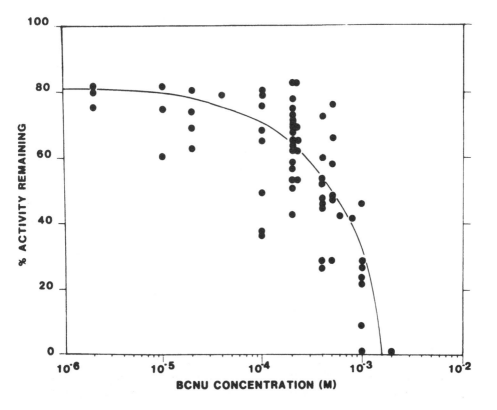

Figure 10.19. Enzyme activity for multiple concentrations of BCNU. The dose that produces 50% inhibition of enzyme activity, I_{50}, can be estimated from data such as these.

10.3.4 Short time–scale kinetics

Each of the various methods described in Section 10.1 for incorporating time into the data base suffers from a considerable disadvantage, which is the finite time taken between mixing cells with substrate or ligand and recording the first event. This "dead time" is due to mixing, location of the containing vessel in the instrument, surge pumping the reaction mixture through the sample feed tube into the flow chamber, and restoration of stable flow. In our instrument the absolute minimum dead time was 15 seconds if everything proceeded smoothly, but more often it was about 20 seconds. There was a clear need to reduce this temporal delay in order to facilitate the analysis of biochemical events taking place over the time period of a few seconds. Such a development would be valuable not only in flow cytoenzymology (e.g. for substrate diffusion kinetics, very rapid–reaction and membrane enzyme analysis), but also for a variety of other biological applications involving very short time scales (e.g. drug uptake and ion flux analysis).

Kachel, Glossner and Schneider (1982) described a flow chamber in which time-resolved measurements could be made within 1 second, but this was not variable. Recently, we have developed a system that markedly reduces the temporal reference frame, allowing it to vary within the interval 1 to 20 seconds (Watson et al. 1988, in press). The technique employs computer-controlled precision-drive syringe pumps, one for substrate and the other for cells, together with variable-length tubing between a mixing chamber and analysis point, selected by an array of zero–dead-space pinch valves. The different tube lengths at a given pump-flow rate give different times between mixing and analysis. A schematic of the system is shown in Figure 10.20.

Calibration was effected using two sets of microbeads with different fluorescence intensities (Polysciences Inc., Warrington, PA, USA). Pumps 1 and 2 were filled with beads of the higher and lower intensity, respectively. The concentrations were adjusted to give a flow rate of 250 beads per second with both pumps running at 100 μl min^{-1}. The concentration of beads in pump 2 (lower intensity) was about 1.4 times greater than in pump 1. Both pumps were activated before data collection in order to fill the input pipes to the mixing chamber, and they were then stopped. The chamber was then flushed through and pump 1 was re-started before data collection. Pump 2, containing the lower-intensity beads, was started during the run after about 1500 of the 10,000 requested events had been recorded. This procedure was repeated for each tube length at various flow rates.

Figure 10.21 shows data for fluorescence versus time as frequency contour plots at the 3-, 6- and 12-event levels, with the fluorescence and flow-rate histograms adjacent to their respective axes. The saw-tooth pattern of the flow-rate histogram is caused by interruption of data collection due to time sharing on the PDP 11/40 computer, which dumps the data to disk, displaying data on the screen between each buffer dump and checking the pump-flow rates before re-commencing data collection. These data were recorded with flow rates of

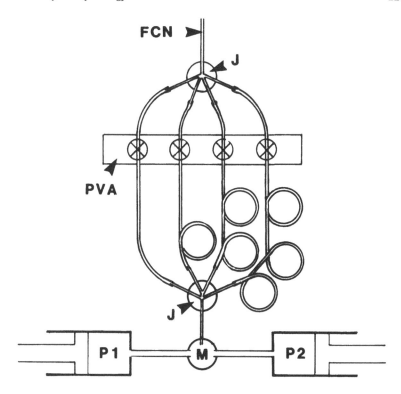

Figure 10.20. Schematic of "stop-flow" device that allows kinetic measurements to be made during the first few seconds of a dynamic reaction.

100 μl min^{-1} from each pump, and panels A, B, C and D (respectively) were obtained for tube lengths of 5, 10, 20 and 40 cm. Full scale on the abscissa is 40 seconds, and each division on the ordinate represents 100 digitisation steps. The long vertical lines drawn through each panel represent the time which is flagged in the data base when pump 2 was activated. The immediate flow-rate increase is apparent, and the sudden surge is also manifest by a widening of the distribution of the higher-intensity beads. The short vertical lines show the time at which the beads from pump 2 first appeared in the data record. The interval between these lines represents the time for the lower-intensity beads from pump 2 to travel from the mixing chamber to the analysis point.

Plots of time from mixing chamber to analysis point versus tube length are shown in Figure 10.22 for pump flow rates of 100, 150, 200 and 250 μl min^{-1}. The regression lines for these pump-flow rates extrapolated almost to a common point, and all the intersections are included in the region defined by the open circle shown in Figure 10.22. Therefore, the four equations defined by the regression analyses were solved simultaneously to give a common intersection point, with X and Y coordinates of -24.73 cm and 0.6 seconds respectively. This value of 24.73 cm represents the volume of the non-variable (dead-space) section of the system, which comprises an individual pair of arms for each four-

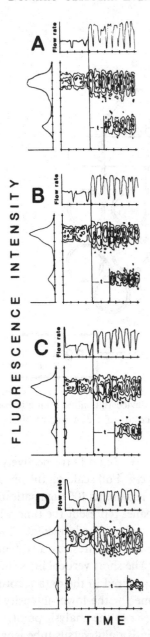

Figure 10.21. Fluorescence-versus-time data contoured at the 3-, 6- and 12-event levels in each panel, with the flourescence and flow-rate histograms adjacent to their respective axes.

way junction, the flow chamber injection needle, the mixing chamber and its outflow. The volume of unit length of the tubing contained in the variable section was measured using Hamilton syringe injection of eight different lengths. The average of the eight readings showed that 1.0 cm was equivalent to 1.046 μl.

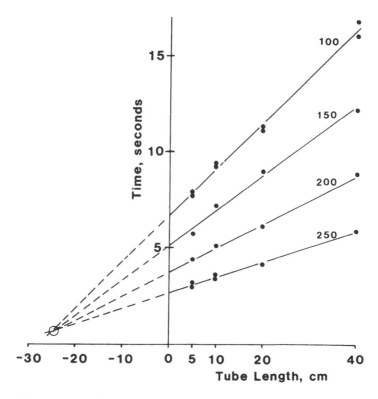

Figure 10.22. Plots of time from mixing chamber to analysis point versus tube length for pump-flow rates of 100, 150, 200 and 250 μl min^{-1}.

The volume of the dead space, excluding the mixing chamber (18.1 μl), was measured (again by Hamilton syringe injection) to be 8.1 μl. Thus the total dead-space volume was 26.2 μl, which corresponds well with the theoretical value of 25.8 μl (24.7 cm × 1.046 μl cm^{-1}) represented by the intersection point in Figure 10.22.

The vertical displacement of the intersection point, 0.6 seconds, constituted a puzzle until we discovered that the assembler-language routines controlling the pump-flow rates, which had been developed for a different application, not only tested but also corrected for any "backlash" before commencing delivery. This involves a feedback hysteresis loop to the computer, and the time taken for backlash correction is inversely proportional to the requested flow rate. At 100 μl min^{-1} the backlash assessment and correction time is between 500 and 600 ms, which accounts for the observed vertical displacement. This minor problem was encountered only when the second pump was started during data collection, that is, when backlash had to be tested for and corrected.

Figure 10.23 shows the slopes of the four regression lines plotted against both the square root and the log of the respective flow rates. Regression analyses gave correlation coefficients of 0.9999 and 0.991 respectively. From this we assumed that the square-root transformation was more appropriate, which gave

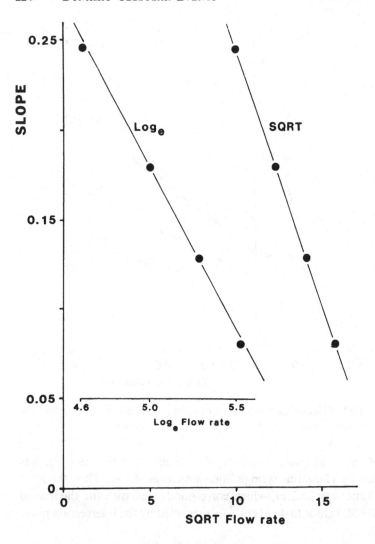

Figure 10.23. Slopes of the four regression lines plotted against both the square root and the log of the respective flow rates. Regression analyses gave correlation coefficients of 0.9999 and 0.991, respectively.

a slope of −0.0283 and intercept of 0.528. From these various data it was possible to express time in relation to flow rate and length of tubing between mixing and analysis points. This is given by the following equation:

$$T = ((0.528 - (0.283 \times \sqrt{F})) \times (L + 24.73)) + 0.6,$$

where T is time, F is the average flow rate of the two pumps and L is tube length. The expression $0.528 - (0.0283 \times \sqrt{F})$, derived from Figure 10.23, is equivalent to the slope m in the equation $y = mx + c$, and $L + 24.73$ is equivalent to x, derived from Figure 10.22. The constant c, 0.6 seconds, can be dropped when both pumps are started before data collection.

The method was tested with EMT6 mouse mammary tumour cells in exponential growth adjusted to a concentration of 2×10^5 cells ml^{-1} and introduced into pump 1 (see Figure 10.20). FDA was made up at a concentration of 2 μM and loaded into pump 2. Both pumps were started, and the high-tension voltage of the green detector was adjusted to record the fluorescein emission histogram with a mean in about channel 600 at pump-flow rates of 100 μl min^{-1} for a tube length of 40 cm. Recordings were then made at each of the four tube lengths at the same flow rate. From the data shown in Figure 10.24 it can be seen that the

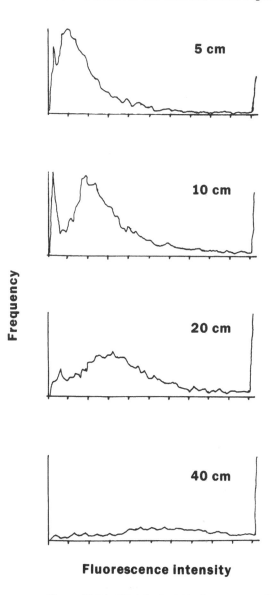

Figure 10.24. FDA hydrolysis fluorescence distributions from EMT6 cells for four tube lengths at the same flow rate.

fluorescence distribution is progressively shifted to higher values with increasing tube length. It should also be noted that the distribution remains constant with time, as the period taken for cells to flow from the mixing chamber to analysis point is constant. This overcomes one of the potential problems associated with continuous-time recording of fast reactions, where – because of the reaction velocity – only relatively few cells can be recorded with a given fluorescence intensity.

A number of additional records were made with various tube lengths and pump-flow rates. The medians of the green fluorescence distributions so obtained are plotted against time in Figure 10.25, which shows that the accumulation of fluorescence was biphasic over the first 16 seconds of the reaction. A linear regression analysis (dashed line) was carried out for the data beyond 5 seconds and this gave a correlation coefficient of 0.99, which is compatible with a linear increase in fluorescein with time between 5 and 16 seconds. However, this is not really describing the biological observation, and a further regression analysis was carried out using the linear transform of the logistic equation (Section 5.2.1);

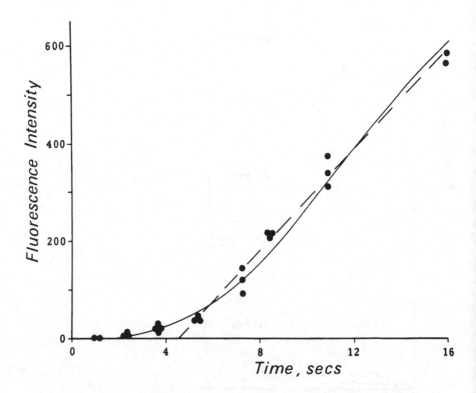

Figure 10.25. Biphasic FDA hydrolysis over the first 16 seconds of the reaction shown by the dashed curve obtained by non-linear regression. The data from 5 to 16 seconds can also be fitted with a straight line (non-interrupted), but this is less likely to represent the biological reality.

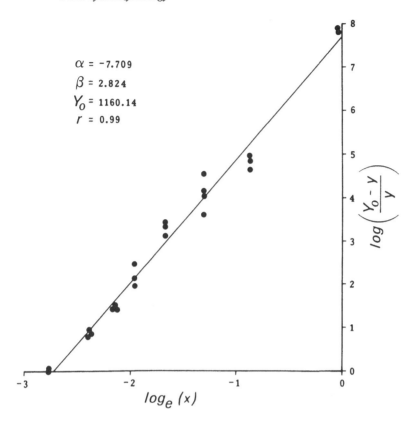

$\alpha = -7.709$
$\beta = 2.824$
$Y_0 = 1160.14$
$r = 0.99$

Figure 10.26. Linear transform analysis of the logistic equation applied to the data in Figure 10.25.

the result of this analysis is shown in Figure 10.26. The initial starting value for Y_0 was 638 fluorescence units, which was incremented until the maximum value for the correlation coefficient was found at 0.99. The final value for Y_0 was 1160 with $\alpha = -7.709$ and $\beta = 2.824$, and the uninterrupted curve in Figure 10.25 was generated with these parameters. As the correlation coefficients were the same in both analyses, the non-linear regression should be preferred because all the data points were included. Furthermore, it is possible to test for the quality of the logistic curve fit to the data, as described in Section 5.2.1 for the fit of the potato data to an inverted exponential. This uses Student's t by comparing the mean deviation D of all the data points from the curve with the standard error of the differences SE_d. As a reminder, these parameters are given by

$$D = \sum d/N;$$
$$SE_d = \sqrt{[(\sum d^2/N) - D^2] \times [N/(N-1)]} \times 1/\sqrt{N}.$$

The data required for this analysis are given in Table 39, where X and Y_{obs} are the experimental data. The regression-model data are given by Y_{expct}, and

Table 39. *Non-linear regression using the logistic curve for the data in Figure 10.25*

X	Y_{obs}	Y_{expct}	Difference	d^2
1.04	0.41	0.59	−0.18	0.03
1.04	0.48	0.59	−0.11	0.01
2.36	8.25	5.87	2.37	5.64
2.36	8.93	5.87	3.06	9.38
2.36	11.00	5.87	5.12	26.26
3.67	11.68	20.08	−8.39	70.46
3.67	18.56	20.08	−1.52	2.31
3.67	19.24	20.08	−0.83	0.70
3.67	30.24	20.08	10.16	103.28
5.33	37.11	55.87	−18.76	351.95
5.33	37.80	55.87	−18.07	326.63
5.33	47.42	55.87	−8.45	71.42
7.27	91.41	125.61	−34.20	1169.50
7.27	124.40	125.61	−1.21	1.46
7.27	145.02	125.61	19.41	376.76
8.49	211.00	183.88	27.12	735.48
8.49	218.56	183.88	34.68	1202.69
8.49	219.93	183.88	36.05	1299.92
10.93	321.65	322.44	−0.79	0.62
10.93	337.46	322.44	15.02	225.64
10.93	375.26	322.44	52.82	2790.16
15.93	569.76	611.53	−41.77	1744.65
15.93	580.07	611.53	−31.46	989.71

the differences between Y_{obs} and Y_{expct} are shown together with the differences squared. Summing the difference and difference squared columns we get 40.09 and 11,504.64 respectively, from which we obtain $D = 1.74$ and $SE_d = 23.31$. The ratio of these two parameters is Student's t, which is 0.359 and represents a highly significant fit, $p > 0.95$. Finally, the logistic fit is also more compatible with the biological reality than is the linear regression from 5 to 16 seconds, and the sigmoidal nature of the data is probably a reflection of the time taken for substrate to diffuse across the external cell membrane and cytoplasm to the site(s) of enzyme action.

10.4   Calcium

Ligand activation of cell-surface receptors linked to the phatidyl inositol pathway generates the second-messenger molecules inositol triphosphate and diacylglycerol (Berridge and Irvine 1984). Inositol triphosphate mobilises calcium, leading to an increase in free calcium ions (Ca^{2+}) in the cytoplasm.

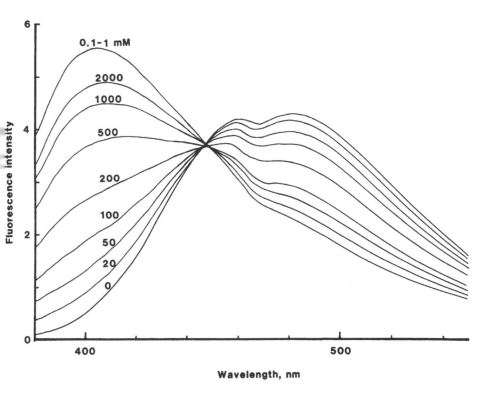

Figure 10.27. Changes in the emission spectrum of indo-1 with increasing Ca^{2+} concentration, given in nM against each curve (except for the maximum, which is mM). These data were redrawn from Grynkiewicz, Poenie and Tsien (1985), who pointed out that the "notch" at 465 nm was an instrument artefact.

A number of Ca^{2+} probes are available, but indo-1 is preferred because it exhibits a shift in emission spectrum on binding calcium. These spectra are shown in Figure 10.27, where it is apparent that as the Ca^{2+} concentration rises there is a progressive shift of the emission to shorter wavelengths. In the absence of Ca^{2+} the emission peak is at 480–485 nm, and at saturating concentrations this has shifted to 400–405 nm. Thus, the ratio of the emissions at 400 nm and 480 nm gives a relative measure of calcium chelated by the probe. Rabinovitch et al. (1986) and more recently Keij et al. (1989) have produced a method for expressing the fluorescence ratio as a concentration of Ca^{2+} by the following equation:

$$nM\ Ca^{2+} = K_d \times \frac{R - R_{min}}{R_{max} - R} \times \frac{S_{f2}}{S_{b2}}.$$

The meanings of these various symbols are as follows. K_d is the indo-1/Ca^{2+} dissociation constant, which is numerically equal to 250 nM (Grynkiewicz, Poenie and Tsien 1985). R is the 400:480 nM fluorescence ratio, and R_{max} and

R_{min} are the 400:480 nM fluorescence ratios for indo-1–loaded cells in the presence of saturating concentrations of Ca^{2+} and in the absence of Ca^{2+}, respectively. A value for R_{max} can be obtained in intact cells by treating with the ionophore ionomycin (Rabinovitch et al. 1986), which allows equilibration of internal Ca^{2+} with the very much higher external concentration. R_{min} is obtained by treating the cells with ionomycin+EGTA. The latter chelates Ca^{2+} external to the cells and hence "drains" Ca^{2+} from the cells. S_{b2} and S_{f2} are the fluorescence values at 480 nm for Ca^{2+}-bound indo-1 and Ca^{2+}-free indo-1 (respectively), and these too are obtained from cells treated respectively with ionomycin and ionomycin+EGTA. A slight variation on this theme was used by Keij et al. (1989), where saponin+EGTA was used instead of ionomycin+EGTA.

Keij et al. (1989) have used the previous formula to present results in the form of Ca^{2+} concentrations in studies involving both T- and B-cell activation with PHA and pneumococcal polysaccharide, respectively. The T-cell results showing [Ca^{2+}] versus time are reproduced in Figure 10.28, where panels A and B give dot plots for each cell and mean population values, respectively. Time was not incorporated directly into the list-mode data base for these studies, but it was "inferred" from the computer-recorded duration of the run as described in Section 10.1.

Indo-1 has also been used to study Ca^{2+} kinetic changes in lymphocytes under various stimuli by Chused et al. (1986, 1987), who were able to distinguish between T and B lymphocytes from mouse spleen cultures, where the mixed population was subjected to B-cell mitogenic stimuli. Within 20 seconds the B-cell component had responded, as indicated by the increase in Ca^{2+}. Indo-1 has also been used to study Ca^{2+} changes in platelets subjected to thrombin stimulation. There was an initial increase in Ca^{2+}, which was maximal by about

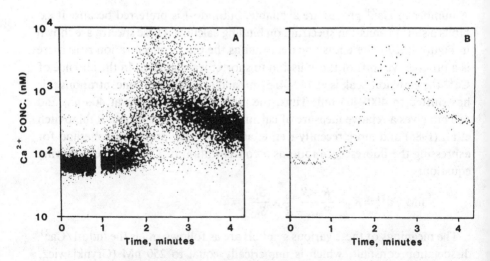

Figure 10.28. T-cell activation showing [Ca^{2+}] versus time, where panels A and B give dot-plots for each cell and population means, respectively.

15 seconds, followed by a slow decline. The use of flow cytometry enabled two subsets to be defined, where one responded fully but the other responded partially or not at all. Rabinovitch et al. (1986) studied not only Ca^{2+} changes during lymphocyte activation but also cell-surface determinants simultaneously with Ca^{2+} using antibodies labelled with phycoerythrin in a multiparameter assay. Heterogeneous Ca^{2+} responses were observed which to some extent related to immunophenotype.

10.5 Membrane transport

Membrane transport and the mechanisms that control molecular traffic across the plasma membrane are of considerable importance in attempts to modulate disease states. This is a very important area, one to which flow cytometry is well suited, but this has not been afforded the attention it either should or could have received. Transport of any agent across the external cell membrane occurs by at least three mechanisms (see the review by Goldenberg and Begleiter 1984): (1) passive diffusion, where the rate of increase of the intracellular concentration is directly proportional to the concentration gradient; (2) facilitated diffusion, where influx involves specific interactions between transport molecules and drug. Unlike passive diffusion, facilitated diffusion is temperature sensitive and exhibits saturation kinetics, but the intracellular concentration cannot exceed the external concentration; (3) active transport, where the agent can be concentrated within the cell against a gradient by using an energy-dependent mechanism. We tend to regard transport as a one-way system from without to within, but clearly these same mechanisms also apply to transport from inside the cell to outside, which has important implications in drug-resistance mechanisms.

Flow cytometric quantitation depends on fluorescence, and any direct measurements of drug uptake must rely on either the drug itself being fluorescent, quenching of a reporter fluorochrome by the drug, or the availability of a fluorescent analogue. Two classes of agents fall into these categories. First, the commonly used anthracyclines, daunorubicin and adriamycin, are both fluorescent and can induce quenching of propidium iodine–DNA emission. Second, fluorescent analogues of the anti-folates, aminopterine and methotrexate, have been synthesised (Gapski et al. 1975; Henderson, Russell and Whiteley 1980; Kumar et al. 1983a,b; Rosowsky et al. 1982). Indirect measurements of intracellular drug content can be made by inhibition kinetics as described in Section 10.3.3.

10.5.1 Anthracyclines

The quenching of propidium iodide–DNA fluorescence as a means of assessing nuclear adriamycin content was demonstrated by Krishan and Ganapathi (1979), who also demonstrated the feasibility of measuring intracellular anthracycline fluorescence directly (Krishan and Ganapathi 1980). Durand

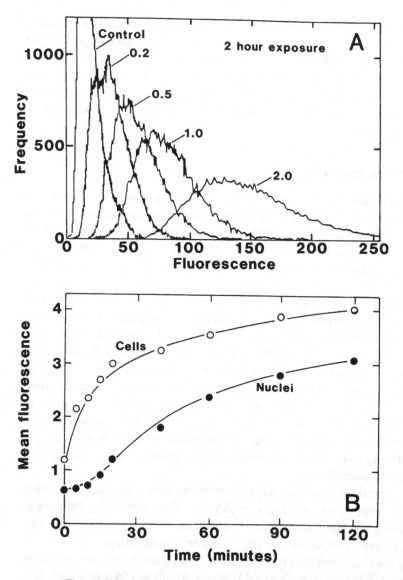

Figure 10.29. Panel A shows adriamycin fluorescence distributions for 2-hour exposures at the concentrations (in μg M⁻¹) indicated adjacent to each histogram. Time courses for the increase in adriamycin fluorescence for whole cells (○) and nuclei (●) exposed to 1.0 μg ml⁻¹ are shown in panel B. Redrawn from Durand and Olive (1981).

and Olive (1981) used similar direct fluorescence techniques to quantitate cellular adriamycin; they also showed time courses for uptake and compared nuclei with whole cells. Some of these data are reproduced in Figure 10.29.

Multiparameter analysis of adriamycin and daunomycin uptake in mouse bone marrow and acute myeloid leukaemia cells has been carried out by Sonneveld and van den Engh (1981) using forward and 90° light scatter properties to identify different subsets (Visser, van den Engh and van Bekkum 1980). Fluorescence was related directly to cytotoxicity, and myeloid progenitor cells were an order of magnitude more sensitive to daunomycin compared with adriamycin, a result that correlated with fluorescence. In contrast, the fluorescence from acute myeloid leukaemia cells was greater for adriamycin than for daunomycin. Further studies of daunomycin fluorescence uptake by rat bone marrow were carried out by Nooter, van den Engh and Sonneveld (1983), where dead cells, lymphocytes, blast cells, granulocytes and red cells were separately identified on light scatter. At higher concentrations the lymphocyte fraction exhibited greater fluorescence than granulocytes, and at given concentrations the dead cells exhibited considerably greater fluorescence than viable cells. These various effects were attributed to quantitative differences in active transport of these agents into different cell types.

Anthracyclines are actively transported both into and out of cells, but by different pump mechanisms. One of the efflux pumps is the 170-kd membrane-associated glycoprotein (gp^{170}) encoded by the multidrug resistance gene MDR-1. Extensive evidence now exists to show that MDR cells have increased numbers of membrane-associated gp^{170} molecules and as a result such cells exhibit lower fluorescence on exposure to anthracyclines compared with sensitive cells, as these agents are actively transported out of MDR cells with greater efficiency. Ross et al. (1988) have used this property to cell sort and isolate a very low-frequency but highly MDR population from within a sensitive P388 cell population on daunorubicin fluorescence. The correct functioning of the efflux pump mechanism can be modulated by Ca^{2+} channel blockers (chlorpromazine, trifluoperazine and verapamil; see Tsuro et al. 1982). A number of studies on the modulation of anthracycline uptake by Ca^{2+} channel blockers have been carried out using flow cytometry in resistant and sensitive cells in both animal and human systems (Krishan et al. 1986, 1987; Morgan et al. 1989; Nooter et al. 1989; Ross, Joneckis and Schiffer 1986; Watson, Morgan and Dive 1989; Watson, Morgan and Smith 1989). The bottom line in all these studies is that the lower accumulation of anthracyclines in MDR cells can be partially reversed by exposure to Ca^{2+} channel blockers that interfere with the efflux pump mechanism.

Some results of such a study by Herweijer, van den Engh and Nooter (1989) are shown in Figure 10.30, where daunorubicin was used as the reporter probe for MDR in sensitive and resistant human ovarian cancer–cell lines. The resistant line, A2780/R, was maintained continuously in medium containing daunorubicin at a concentration of 2 μM, but the cells were incubated in drug-free medium for 48 hours prior to analysis. Time courses for the uptake of daunorubicin are shown for the sensitive (A2780/S) and resistant (A2780/R) lines, and the former shows considerably greater uptake. These results are interesting not only for the effect they show but also for the information that can be obtained. It is clear that these uptake curves are of the inverted exponential type,

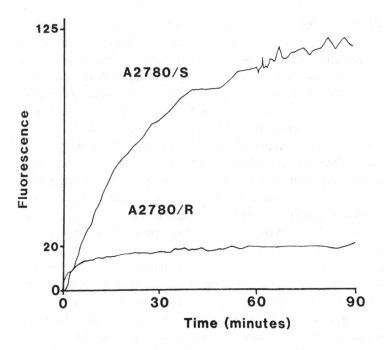

Figure 10.30. Daunorubicin time course uptake in sensitive (A2780/S) and resistant (A2780/R) cell lines. Redrawn from Herweijer, van den Engh and Nooter (1989).

which suggests that the membrane traffic can be approximated with first-order processes; a schematic for these is shown in Figure 10.31. The rate equation is

$$\frac{\delta S}{\delta t} = k_1 E - k_2 S,$$

where S represents the intracellular drug concentration in the sensitive cells and E represents the extracellular concentration. The respective rate constants are k_1 for influx and k_2 for efflux. This equation can be rearranged to give

$$\frac{\delta S}{\delta t} + k_2 S = k_1 E.$$

On integration this equation yields the familiar inverted exponential

$$S = \frac{k_1}{k_2} E (1 - e^{-k_2 t}),$$

where k_2 is the rate constant for efflux and where the uptake curve should reach an asymptote, Y_0, of $E(k_1/k_2)$. (The derivation of this equation is given in Appendix 9.) The comparable equation for resistant cells is

$$R = \frac{k_3}{k_4} E (1 - e^{-k_4 t}),$$

Sensitive

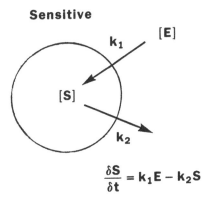

$$\frac{\delta S}{\delta t} = k_1 E - k_2 S$$

Figure 10.31. Simple schematic for influx and efflux of daunorubicin.

where R is the intracellular drug concentration, k_3 is the rate constant for influx, k_4 is the rate constant for efflux and where the uptake curve should reach an asymptote of $E(k_3/k_4)$.

The control daunorubicin uptake results of Herweijer et al. (1989) shown in Figure 10.30 were analysed (it's always more fun "having a go" at someone else's data) using the linear transform of the inverted exponential (Section 5.2.1). The results are given in Table 40. The fit between the mathematical model and the data was good as can be seen from the values for Student's t and p in Table 40 and in Figure 10.32, where the curves are the inverted exponentials and the symbols are "my" data points measured from the figures in the authors' publication.

We can now see that the rate constant for efflux across the external membrane was about six-fold greater in MDR cells compared with the sensitive parent. However, what about the influx rate constants k_1 and k_3? We can see from the uptake equations that $Y_0 = (k_1/k_2)E$ for the sensitive cells and $Y_0 = (k_3/k4)E$ for the resistant ones. Thus,

$$k_1 E = Y_0 \times k_2 = 124.4 \times 0.0365 = 4.54 \text{ (sensitive)};$$

$$k_3 E = Y_0 \times k_4 = 20.0 \times 0.2091 = 4.18 \text{ (resistant)}.$$

Table 40. *Asymptotes and efflux-rate constants for the sensitive and resistant cell lines with Student's t and p values*

Cells	Y_0	Efflux-rate constant	t	p
Sensitive	124.4	0.0365	0.32	>0.60
Resistant	20.0	0.2091	2.12	>0.05

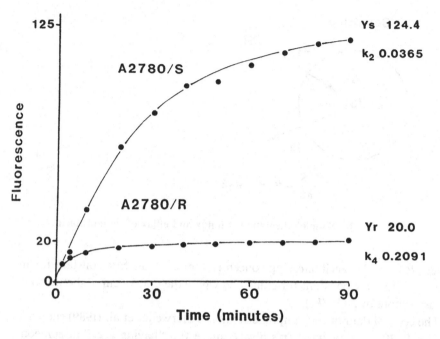

Figure 10.32. Non-linear curve fitting of the inverted exponential to the control daunorubicin uptake data of Herweijer et al. (1989).

These two values, considering the potential inaccuracies of my measurements from the original figures (particularly after enlargement on the photocopier), are remarkably close: within 8%. Thus, we can reasonably assume that the influx-rate constant is the same in both cells and that the difference in fluorescence levels is due entirely to greater efflux in the resistant cells. It is also clear that if we can relate the fluorescence units of the Y_0 values to the extracellular drug concentration units E, then we can also obtain the rate constant(s) for drug influx.

These conclusions were substantiated in the work of Herweijer et al. (1989) with the calcium channel blockers verapamil and cyclosporin A, which inhibit gp^{170}. After addition of these agents to the resistant cells there was an immediate dose-dependent increase in daunorubicin accumulation which followed first-order kinetics, but there was no comparable change in the sensitive cells.

10.5.2 Methotrexate

Similar types of studies have been carried out with cells sensitive and resistant to the antimetabolite methotrexate using fluorescent analogues of the latter. However, methotrexate resistance is different from that in MDR, and two of the three mechanisms identified are attributable to Fisher (1961, 1962). The enzyme dihydrofolate reductase (DHFR) converts folic acid to folinic acid, which is required in one of the pathways to DNA synthesis. Methotrexate binds to DHFR with an affinity about 40-fold greater than that of folic acid and blocks its action. Fisher (1961) demonstrated that methotrexate resistance could be caused by elevation of dihydrofolate reductase levels. Cells grown continuously in medium containing methotrexate exhibit amplification of the DHFR gene (Alt, Kellems and Schimke 1976; Hanggi and Littlefield 1976), where the multiple copies are contained in "double minute" chromosomes and are manifest in such cells at metaphase. Stable mutants are obtained when these double minutes become integrated into the normal chromosomes and the amplified gene can then be transmitted predictably from one generation to the next; this confers considerable methotrexate resistance on the population. Fisher (1962) also demonstrated that the rate of influx of methotrexate (MTX) was about 15-fold lower in resistant than in sensitive cells. At equilibrium the intracellular MTX concentration was between 10- and 30-fold lower in the resistant cells. However, this was not due to metabolic changes, as dihydrofolate reductase levels were equal in both the resistant and sensitive cells. Changes in methotrexate membrane transport contributing to resistance (Schimke 1984) have also been demonstrated by Sirontak et al. (1981), and Dembo, Sirontak and Moccia (1984) have presented evidence that the cellular influx and efflux pathways for methotrexate are different. The third mechanism of methotrexate resistance is due to an alteration of the affinity of methotrexate binding to DHFR (Flintoff, Davidson and Siminovitch 1976; Haber et al. 1981).

To date, relatively limited work has been carried out on methotrexate-resistant cells using flow cytometry. This has been due in part to the difficulty of producing fluorescent analogues with low background, as pointed out by Gaudray, Trotter and Wahl (1986). However, these authors overcame this problem with a fluorescenated derivative of the ligand, and presented data which suggested that staining heterogeneity with the analogue was partly due to transport variability within the population. Assaraf and Schimke (1987) and Assaraf, Seamer and Schimke (1989) loaded cells with fluorescent methotrexate and then examined the ability of hydrophilic and lipophilic antifolates to displace the probe. Methotrexate-resistant cells incubated with methotrexate (hydrophilic) exhibited no reduction in fluorescence compared with sensitive cells, but incubation with lipophilic analogues produced reductions in fluorescence in both sensitive and resistant cells. This is very compelling evidence for a membrane active-transport mechanism for methotrexate and hydrophilic analogues which is deficient in resistant cells. The lipophilic analogues enter cells by passive diffusion without the need for an active-transport mechanism, so they displaced the fluorescenated probe bound to DHFR in both the sensitive and resistant cell lines. Similar types of studies have been carried out by Rosowsky et al. (1986) and Wright et al. (1987). Figure 10.33 shows the uptake of fluorescent methotrexate versus time in a parental chinese hamster cell line, AA8, and in a DHFR-deficient cell line, DG-44. The former exhibits considerably greater fluorescence which is attained at a more rapid rate. The effect of a temperature reduction, which

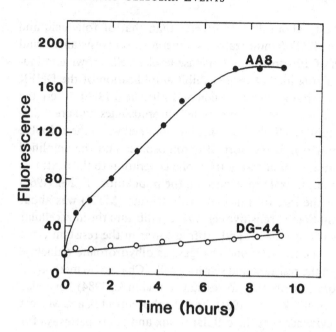

Figure 10.33. Uptake kinetics for fluorescent methotrexate in a parental cell line (AA8, O) and a dihydrofolate reductase–deficient cell line (DG-44, O) (redrawn from Assaraf et al. 1989).

slows the uptake rate of active transport processes but not of passive diffusion, is shown in Figure 10.34. There was a profound decrease in fluorescent methotrexate uptake in parental CHO when the temperature was decreased from 37°C to 4°C, an effect easily observed in the initial stages of uptake with a temperature decrease to 27°C. Both of these last two data sets were redrawn from Assaraf et al. (1989).

10.5.3 Chloroethylnitrosoureas

Passive diffusion is the means by which the nitrosoureas CCNU, BCNU (Begleiter, Lam and Goldenberg 1977) and chlorozotocin (Lam, Talgoy and Goldenberg 1980) gain access to the cell interior, as the uptake of each is unsaturable, independent of temperature and unaltered by metabolic inhibitors. It was shown in Section 10.3.3 that esterase inhibition by carbamoylation with nitrosoureas and related isocyanates can be assayed flow cytometrically and comparisons can be made between agents by using the I_{50} value, which is the dose required to induce 50% inhibition of activity. An example of BCNU inhibition was given in that section, and similar assays were performed with a number of drugs using both flow cytometry for intact cells and conventional spectrofluorimetry with sonicated preparations. The I_{50} values for the two methods are compared in Table 41, together with the ratio of results from intact cells to those in sonicates.

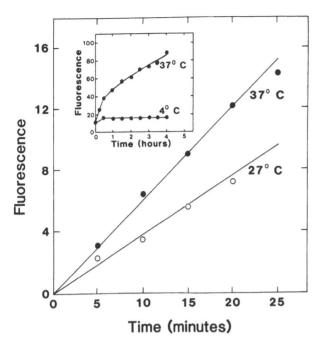

Figure 10.34. Temperature-dependent uptake of fluorescent methotrexate in parental CHO cells. The insert shows that cooling from 37°C to 4°C effectively abolishes uptake. The main graph shows that a reduction in uptake can be measured with a temperature decrease to only 27°C (redrawn from Assaraf et al. 1989).

Table 41. I_{50} *values for a variety of nitrosoureas and related isocyanates*

Drug	I_{50} values (M) Intact cells	Sonicates	Ratio
CHI	1.1×10^{-5}	2.6×10^{-5}	0.423
CCNU	3.8×10^{-5}	8.3×10^{-5}	0.458
BCNU	5.0×10^{-5}	9.2×10^{-5}	0.543
CEI	1.0×10^{-5}	1.6×10^{-5}	0.625
TCNU	1.8×10^{-4}	9.9×10^{-5}	1.818
ACNU	7.3×10^{-3}	3.2×10^{-3}	2.281
CHLOZ	$>1.5 \times 10^{-2}$	2.4×10^{-3}	>6.250
GANU	$>1.5 \times 10^{-2}$	2.4×10^{-3}	>6.250

The results in this table are interesting. The I_{50} ratios for CHI, CCNU, BCNU and CEI indicate that only about half the concentration of drug is required to obtain 50% esterase activity inhibition in whole cells compared with sonicates. With TCNU and ACNU this is reversed, with double the drug concentration required in whole cells compared with sonicates, and with chlorozotozin and GANU this has increased more than six-fold. These results almost exactly parallel the hydrophilicity of the agents. CHI, CCNU, BCNU and CEI are lipophilic, but chlorozotozin and GANU are hydrophilic and readily soluble in water at physiological pH and temperature, and would not be expected to cross the external cell membrane without an active-transport mechanism. We have seen previously that nitrosoureas enter cells by passive diffusion, so we would not expect chlorozotozin and GANU to enter cells. This conclusion would seem to be substantiated by the data in Table 41, and the method seems capable of assessing not only the ability of such agents to cross the external cell membrane, but also their capability of inducing damage having entered the cell.

11

Multivariate analysis primer

This brief chapter is merely intended to introduce the idea of multivariate analysis, which will not be pursued in detail as the topic is appropriate to a more advanced book.

11.1 Multivariate density function

We saw in Section 2.5.2 that the 1-dimensional normal density curve is given by the following formula:

$$y = \frac{1}{\sigma\sqrt{2\pi}} \exp\left[-\frac{1}{2}\left(\frac{x-\bar{x}}{\sigma}\right)^2\right].$$

Just as the concept of calculating the absolute distance between points in n-dimensional hyperspace was developed from the 1-dimensional case (Section 2.2), so too can the n-dimensional normal density function be developed from the 1-dimensional Gaussian as follows:

$$\phi(x_1, \ldots, x_n) = \frac{1}{(2\pi)^{n/2}(\sigma_1, \ldots, \sigma_n)} \exp\left[-\frac{1}{2}\sum_{i=1}^{i=n}\left(\frac{x_i-\bar{x}_i}{\sigma_i}\right)^2\right],$$

where $\bar{x}_i$ and σ_i are the mean and standard deviation of the ith distribution. This is the density function, which forms the basis for multivariate analysis.

The variates $x_1 x_2, \ldots, x_n$ and their means $\bar{x}_1, \bar{x}_2, \ldots, \bar{x}_2$ can be represented in matrix notation, where

$$\Xi' = [x_1, x_2, \ldots, x_n] \quad \text{and} \quad \Upsilon' = [\bar{x}_1, \bar{x}_2, \ldots, \bar{x}_n]$$

which now represent vectors of the variates and the means. The variance component in the exponent now becomes a diagonal variance matrix Σ^2, where

$$\Sigma^2 = \begin{bmatrix} \sigma_1^2 & \ldots & 0 \\ & \ddots & \\ 0 & \ldots & \sigma_n^2 \end{bmatrix},$$

and the density function can now be written as

$$\phi(x) = \frac{1}{(2\pi)^{n/2}|\Sigma|} \exp\left[-\frac{1}{2}(\Xi - \Upsilon)'\Sigma^{-2}(\Xi - \Upsilon)\right].$$

11.2   Correlated bivariate density function

The density function shown in the preceding equation applies if the variates are independent and *not* correlated. However, this possibility can be incorporated by modifying the variance matrix Σ^2 to include the correlation coefficients r, where r_{ij} is the correlation between variates x_i and x_j. In a joint correlated bivariate distribution, $n = 2$ and

$$\Sigma^2 = \begin{bmatrix} \sigma_1^2 & r\sigma_1\sigma_2 \\ r\sigma_1\sigma_2 & \sigma_2^2 \end{bmatrix},$$

which is now a variance–covariance matrix; the joint density function becomes

$$\phi(x_1, x_2) = \frac{1}{2\pi\sigma_1\sigma_2\sqrt{1-r^2}} \exp\left\{-\frac{1}{2}\frac{1}{1-r^2}\left[\left(\frac{x_1-\bar{x}_1}{\sigma_1}\right)^2 \right.\right.$$
$$\left.\left. -2r\left(\frac{x_1-\bar{x}_1}{\sigma_1}\frac{x_2-\bar{x}_2}{\sigma_2}\right) + \left(\frac{x_2-\bar{x}_2}{\sigma_2}\right)^2\right]\right\}.$$

We can see immediately that if the variates x_1 and x_2 are not correlated (i.e. if $r = 0$), then the correlation component in the exponent becomes zero and the $1 - r^2$ terms become unity. Also, the variance–covariance matrix reverts to a diagonal variance matrix, and the joint density function reduces to

$$\phi(x_1, x_2) = \frac{1}{2\pi\sigma_1\sigma_2} \exp\left\{-\frac{1}{2}\left[\left(\frac{x_1-\bar{x}_1}{\sigma_1}\right)^2 + \left(\frac{x_2-\bar{x}_2}{\sigma_2}\right)^2\right]\right\},$$

exactly the form in which this function was first encountered.

The correlated joint density function can be simplified with z-transformations in a manner similar to that for the 1-dimensional function (see Section 2.5.2). If we replace $(x_1-\bar{x}_1)/\sigma_1$ by z_1 and $(x_2-\bar{x}_2)/\sigma_2$ by z_2 then we obtain the standardised form, which can be used as a bivariate probability function:

$$\phi(x_1, x_2) = \frac{1}{2\pi\sqrt{1-r^2}} \exp\left[-\frac{1}{2}\frac{1}{1-r^2}(z_1^2 - 2rz_1z_2 + z_2^2)\right].$$

11.3   2-dimensional surface fitting

The various functions just outlined can be used for any 2-dimensional data set where the peaks in the data conform approximately to a bivariate normal distribution. Dean, Kolla and van Dilla (1989) have applied these functions to the analysis of bivariate flow karyotypes using the Hoechst–chromomycin A3 staining method, which assays for AT:GC ratios. Some of their data are reproduced in Figure 11.1, where the top two panels show a perspective view

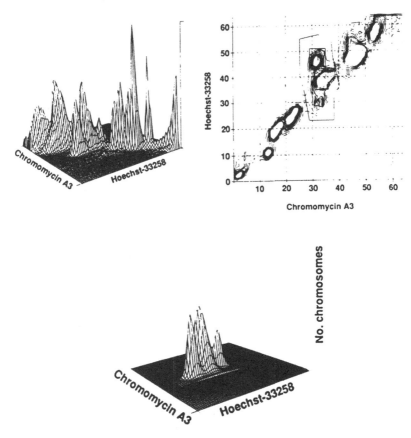

Figure 11.1. Bivariate flow karyotype; perspective view (top left) and contour plot (top right). The bottom panel is the surface-fitted distribution to the data in the polygon on the top right panel.

of the data (left) and the corresponding contour plot (right). A polygon surrounds three of the peaks in the latter; these were surface fitted to produce the bottom panel.

We are using the bivariate functions for two purposes in the development of our sequential automated data-analysis system. First, the distribution is used to check for differences between successive data sets; this works as follows. The top left panel of Figure 11.2, PA, shows 90° light scatter plotted on the ordinate versus DNA on the abscissa, with the associated 1-dimensional histograms adjacent to the respective axes. There are three peaks on the DNA axis – corresponding to G1, mid S phase–arrested and G2+M cells – and three elliptical gates were set on these populations. The individual gated populations with the numbers in each are shown in the remaining three panels where P1, P2 and P3 correspond to the G1, mid S phase–arrested and G2+M cells, respectively. A regression analysis of y on x has been carried out on the data in each of these

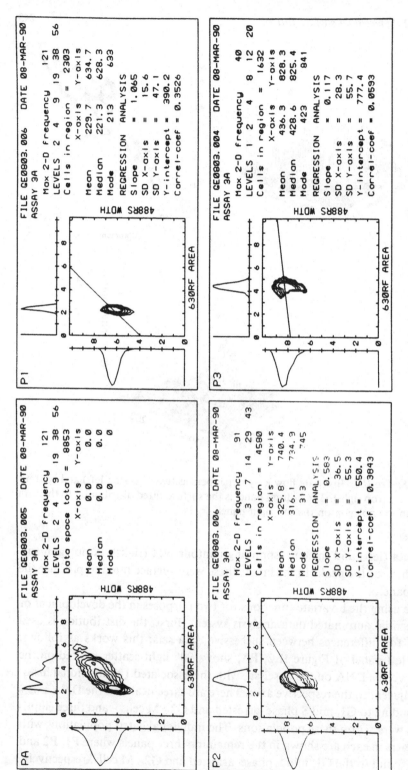

Figure 11.2. Light scatter versus DNA (top left). Panels P1 (top right), P2 (bottom left) and P3 (bottom right) correspond to G1, mid S phase-arrested and G2+M gated cells. A *y*-on-*x* regression has been carried out for each population.

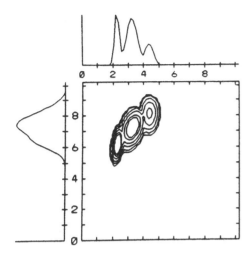

Figure 11.3. "Reconstruction" of the whole data set from the various parameters shown in Figure 11.2, using the bivariate joint density function.

panels to obtain the slopes, standard deviations, Y-intercepts and correlation coefficients, which are also displayed on each panel. We can see that these data are all that are required for input to the bivariate joint density function, which is scaled for the numbers of cells in each gate. The resulting summation of the three distributions so generated is shown in Figure 11.3. This is recognisably similar to the top left panel of Figure 11.2, and the sum of squared differences between the two surfaces is recorded to act as a baseline. If the next data set encountered in the sequential analysis exhibits a sum of squared differences that is statistically different from the baseline, it is interpreted as a new data set, which now needs to be gated differently. Cluster-analysis algorithms are invoked which use the bivariate probability function in the decision process to define clusters.

12

Epilogue

Mathematicians and statisticians will have found this book very elementary, whilst completely non-mathematically and non-statistically oriented biologists will probably have found some of this a little heavy going. Nevertheless, it was written for the latter group and I hope that it proves useful even if you *have* found it heavy going.

Flow cytometry instrumentation is becoming not only increasingly sophisticated but also more generally available, which enables more of us to perform more experiments with increasing sophistication. Analysis and interpretation of the data is the most important single aspect in flow cytometry, for without it the data remain meaningless. Mathematics and statistics are the tools required for that analysis and interpretation; non-mathematically and non-statistically oriented biologists increasingly will have recourse to these disciplines, into which they must acquire at least some insight. If this book helps a single person to obtain such insight then I will regard the effort involved in writing it as worthwhile.

Mathematics and statistics are also an integral part of all our lives in ways we might be unaware of and hence cannot appreciate. As an illustration I'll return to the undergarment problem mentioned at the outset (I hadn't forgotten). The various measurements were resolved into four basic categories, which can be summarised as circumference, vertical displacement, separation and volume. Circumference was well correlated with vertical displacement, $r \simeq 0.8$. This is not too surprising as larger/smaller people tend to be "symmetrically" larger/smaller, in the sense that if circumference is large/small then any vertical section through the plane of that circumference will be correspondingly large/small. In other words, we all have the same overall shape irrespective of size and gender. Separation was poorly correlated with not only circumference, $r \simeq 0.3$, but also vertical displacement, $r \simeq 0.2$. Volume was more poorly correlated with both circumference and vertical displacement, $r \simeq 0.15$ and $r \simeq 0.1$ respectively, but it was reasonably well correlated with separation, $r \simeq 0.6$. We can surmise from these results that the volume component of these various measurements was a variable essentially independent of circumference and vertical displacement and hence of underlying shape.

The manufacturing problem was finally resolved into two components. Circumference plus vertical displacement were treated as a single variable (as $r \simeq 0.8$) versus volume/separation. Both sets of measurements were skewed to the right, so log conversions were carried out. The distributions were then split into six groups:

$$< -2\sigma$$
$$-2\sigma \Rightarrow -1\sigma$$
$$-1\sigma \Rightarrow 0\sigma$$
$$0\sigma \Rightarrow +1\sigma$$
$$+1\sigma \Rightarrow 2\sigma$$
$$> +2\sigma.$$

Hence, there was a total of 36 different size/shapes, and the numbers of each size/shape manufactured were directly proportional to the respective volumes under the bivariate distribution.

A number of other problems of detail also needed to be resolved. As the circumference distribution was log converted, there was a greater range from (say) $+1\sigma \Rightarrow +2\sigma$ than from $-2\sigma \Rightarrow -1\sigma$. This was solved by using an elastic insert whose length was proportional to the σ interval. Vertical-displacement variation within each group was taken care of by using manually adjustable non-elasticised material. Volume variation within each group was not a great problem as the enclosed anatomical features are semi-fluid; this component of the article was constructed with a partially elasticised material. Separation variation was solved with a small elastic insert between the volume-containing components. The final design feature was the addition of a few "frilly" bits here and there which had absolutely nothing to do with function. This was the "appeal" factor, recognised as the most important (yet least understood) single selling feature; this has not yet yielded to statistical analysis. Before I met the lady responsible for all these measurements and statistics I had absolutely no idea that such care and attention were invested in the design and construction of clothing, did you?

Appendix 1
Numerical integrating routine

```
C   ...Micro VAX-3 F77 version 15-May-90 **********************************
    REAL FUNCTION ERFX(X)
C   Algorithm 209 (D. Ibbertson).   COMM. A.C.M. OCTOBER 1963
C   Calculates the AREA under the normal curve from -INF to X
C   In call statement X=(X-XBAR)/SD
    IF (X.EQ.0.) GO TO 1
    Y=X*.5
    IF (Y.LT.0.) Y=-Y
    IF (Y.GE.3.) GO TO 2
    IF (Y.GE.1.) GO TO 3
    W=Y**2
    Z=((0.000124818987*W-0.001075204047)*W+0.005198775019)*W
    Z=(((Z-0.019198292004)*W+0.059054035642)*W-0.151968751364)*W
    Z=(((Z+0.319152932694)*W-0.531923007300)*W+0.797884560593)*Y*2.
4   IF (X.GE.0.) GO TO 5
    Z=(1.-Z)*.5
    GO TO 6
5   Z=(1.+Z)*.5
6   ERFX=Z
    RETURN
3   Y=Y-2.
    Z=((-0.000045255659*Y+0.000152529290)*Y-0.000019538132)
    Z=(((Z*Y-0.000676904986)*Y+0.001390604284)*Y-0.000794620820)*Y
    Z=(((Z-0.002034254874)*Y+0.006549791214)*Y-0.010557625006)*Y
    Z=(((Z+0.011630447319)*Y-0.009279453341)*Y+0.005353579108)*Y
    Z=((Z-0.002141268741)*Y+0.000535310849)*Y+0.999936657524
    GO TO 4
2   Z=1.
    GO TO 4
1   Z=0.
    GO TO 4
    END
```

Appendix 2
Normal distribution probabilities

The normal distribution is symmetrical, so only half is represented here. z is the number of standard deviations from the mean, and the values in the table are the areas under the normal distribution "outside" the value of z. The left column gives z to one decimal place and the top row gives the second decimal place. We find 0.2946 for $z = 0.54$, thus a fraction 0.2946 of the normal curve area is beyond 0.54 standard deviations from the mean. This applies both to the left (−) and the right (+) of the mean.

z	.00	.01	.02	.03	.04	.05	.06	.07	.08	.09
0.0	.5000	.4960	.4920	.4880	.4840	.4801	.4761	.4721	.4681	.4641
0.1	.4602	.4562	.4522	.4483	.4443	.4404	.4364	.4325	.4286	.4247
0.2	.4207	.4168	.4129	.4090	.4052	.4013	.3974	.3936	.3879	.3859
0.3	.3821	.3783	.3745	.3707	.3669	.3632	.3594	.3557	.3520	.3483
0.4	.3446	.3409	.3372	.3336	.3300	.3264	.3228	.3192	.3156	.3121
0.5	.3085	.3050	.3015	.2981	.2946	.2912	.2877	.2843	.2810	.2776
0.6	.2743	.2709	.2676	.2643	.2611	.2587	.2546	.2514	.2483	.2451
0.7	.2420	.2389	.2358	.2327	.2296	.2266	.2236	.2206	.2177	.2148
0.8	.2119	.2090	.2061	.2033	.2005	.1977	.1949	.1921	.1894	.1867
0.9	.1841	.1814	.1788	.1762	.1736	.1711	.1685	.1660	.1635	.1611
1.0	.1587	.1562	.1539	.1515	.1492	.1469	.1446	.1423	.1401	.1379
1.1	.1357	.1335	.1314	.1292	.1271	.1251	.1230	.1210	.1190	.1170
1.2	.1151	.1131	.1112	.1093	.1075	.1056	.1038	.1020	.1003	.0985
1.3	.0968	.0951	.0934	.0918	.0901	.0885	.0869	.0853	.0838	.0823
1.4	.0808	.0793	.0778	.0764	.0749	.0735	.0721	.0708	.0694	.0681
1.5	.0668	.0655	.0643	.0630	.0618	.0606	.0594	.0582	.0557	.0559
1.6	.0549	.0537	.0526	.0516	.0505	.0495	.0485	.0475	.0465	.0455
1.7	.0446	.0436	.0427	.0418	.0409	.0401	.0392	.0384	.0375	.0367
1.8	.0359	.0351	.0344	.0336	.0329	.0322	.0314	.0307	.0301	.0294
1.9	.0287	.0281	.0274	.0268	.0262	.0256	.0250	.0244	.0239	.0233
2.0	.0228	.0222	.0217	.0212	.0207	.0201	.0197	.0192	.0188	.0183
2.1	.0179	.0174	.0170	.0166	.0162	.0158	.0154	.0150	.0146	.0143
2.2	.0139	.0136	.0132	.0129	.0125	.0122	.0119	.0116	.0113	.0110
2.3	.0107	.0104	.0102	.0099	.0096	.0094	.0091	.0089	.0087	.0084
2.4	.0082	.0080	.0078	.0075	.0073	.0071	.0069	.0068	.0066	.0064

z	.00	.01	.02	.03	.04	.05	.06	.07	.08	.09
2.5	.0062	.0060	.0059	.0057	.0055	.0054	.0052	.0051	.0049	.0048
2.6	.0047	.0045	.0044	.0043	.0041	.0040	.0039	.0038	.0037	.0036
2.7	.0035	.0034	.0033	.0032	.0031	.0030	.0029	.0028	.0027	.0026
2.8	.0026	.0025	.0024	.0023	.0022	.0021	.0021	.0020	.0019	.0019
2.9	.0019	.0018	.0018	.0017	.0016	.0016	.0015	.0015	.0014	.0014
3.0	.0013	.0013	.0013	.0012	.0012	.0011	.0011	.0011	.0010	.0010
3.1	.0010	.0009	.0009	.0009	.0008	.0008	.0008	.0008	.0007	.0007

Appendix 3
Variance ratio tables

In each of these tables, n_1 and n_2 are the number of degrees of freedom in the greater and lesser variance estimates, respectively.

Variance ratios for $p = 0.10$

	n_1								
n_2	1	2	3	4	5	6	12	24	∞
1	39.86	49.50	53.59	55.83	57.24	58.20	60.70	62.00	63.33
2	8.53	9.00	9.16	9.24	9.29	9.33	9.41	9.45	9.49
3	5.54	5.46	5.39	5.34	5.31	5.28	5.22	5.18	5.13
4	4.54	4.32	4.19	4.11	4.05	4.01	3.90	3.83	3.76
5	4.06	3.78	3.62	3.52	3.45	3.40	3.27	3.19	3.10
6	3.78	3.46	3.29	3.18	3.11	3.05	2.90	2.82	2.72
7	3.59	3.26	3.07	2.96	2.88	2.83	2.67	2.58	2.47
8	3.46	3.11	2.92	2.81	2.73	2.67	2.50	2.40	2.29
9	3.36	3.01	2.81	2.69	2.61	2.55	2.38	2.28	2.16
10	3.28	2.92	2.73	2.61	2.52	2.46	2.28	2.18	2.06
11	3.23	2.86	2.66	2.54	2.45	2.39	2.21	2.10	1.97
12	3.18	2.81	2.61	2.48	2.39	2.33	2.15	2.04	1.90
13	3.14	2.76	2.56	2.43	2.35	2.28	2.10	1.98	1.85
14	3.10	2.73	2.52	2.39	2.31	2.24	2.05	1.94	1.80
15	3.07	2.70	2.49	2.36	2.27	2.21	2.02	1.90	1.76
16	3.05	2.67	2.46	2.33	2.24	2.18	1.99	1.87	1.72
17	3.03	2.64	2.44	2.31	2.22	2.15	1.96	1.84	1.69
18	3.01	2.62	2.42	2.29	2.20	2.13	1.93	1.81	1.66
19	2.99	2.61	2.40	2.27	2.18	2.11	1.91	1.79	1.63
20	2.97	2.59	2.38	2.25	2.16	2.09	1.89	1.77	1.61
21	2.96	2.57	2.36	2.23	2.14	2.08	1.88	1.75	1.59
22	2.95	2.56	2.35	2.22	2.13	2.06	1.86	1.73	1.57
23	2.94	2.55	2.34	2.21	2.11	2.05	1.84	1.72	1.55
24	2.93	2.54	2.33	2.19	2.10	2.04	1.83	1.70	1.53
25	2.92	2.53	2.32	2.18	2.09	2.02	1.82	1.69	1.52
26	2.91	2.52	2.31	2.17	2.08	2.01	1.81	1.68	1.50
27	2.90	2.51	2.30	2.17	2.07	2.00	1.80	1.67	1.49
28	2.89	2.50	2.29	2.16	2.06	2.00	1.79	1.66	1.48

Variance ratios for $p = 0.10$ (cont.)

n_2	n_1 1	2	3	4	5	6	12	24	∞
29	2.89	2.50	2.28	2.15	2.06	1.99	1.78	1.65	1.47
30	2.88	2.49	2.28	2.14	2.05	1.98	1.77	1.64	1.46
40	2.84	2.44	2.23	2.09	2.00	1.93	1.71	1.57	1.38
60	2.79	2.39	2.18	2.04	1.95	1.87	1.66	1.51	1.29
120	2.75	2.35	2.13	1.99	1.90	1.82	1.60	1.45	1.19
∞	2.71	2.30	2.08	1.94	1.85	1.77	1.55	1.38	1.00

Variance ratios for $p = 0.05$

n_2	n_1 1	2	3	4	5	6	12	24	∞
1	161.4	199.5	215.7	224.6	230.2	234.0	243.9	249.0	254.3
2	18.51	19.00	19.16	19.25	19.30	19.33	19.41	19.45	19.50
3	10.13	9.55	9.28	9.12	9.01	8.94	8.74	8.64	8.53
4	7.71	6.94	6.59	6.39	6.26	6.16	5.91	5.77	5.63
5	6.61	5.79	5.41	5.19	5.05	4.95	4.68	4.53	4.36
6	5.99	5.14	4.76	4.53	4.39	4.28	4.00	3.84	3.67
7	5.59	4.74	4.35	4.12	3.97	3.87	3.57	3.41	3.23
8	5.32	4.46	4.07	3.84	3.69	3.58	3.28	3.12	2.93
9	5.12	4.26	3.86	3.63	3.48	3.37	3.07	2.90	2.71
10	4.96	4.10	3.71	3.48	3.33	3.22	2.91	2.74	2.54
11	4.84	3.98	3.59	3.36	3.20	3.09	2.79	2.61	2.40
12	4.75	3.88	3.49	3.26	3.11	3.00	2.69	2.50	2.30
13	4.67	3.80	3.41	3.18	3.02	2.92	2.60	2.42	2.21
14	4.60	3.74	3.34	3.11	2.96	2.85	2.53	2.35	2.13
15	4.54	3.68	3.29	3.06	2.90	2.79	2.48	2.29	2.07
16	4.49	3.63	3.24	3.01	2.85	2.74	2.42	2.24	2.01
17	4.45	3.59	3.20	2.96	2.81	2.70	2.38	2.19	1.96
18	4.41	3.55	3.16	2.93	2.77	2.66	2.34	2.15	1.92
19	4.38	3.52	3.13	2.90	2.74	2.63	2.31	2.11	1.88
20	4.35	3.49	3.10	2.87	2.71	2.60	2.28	2.08	1.84
21	4.32	3.47	3.07	2.84	2.68	2.57	2.25	2.05	1.81
22	4.30	3.44	3.05	2.82	2.66	2.55	2.23	2.03	1.78
23	4.28	3.42	3.03	2.80	2.64	2.53	2.20	2.00	1.76
24	4.26	3.40	3.01	2.78	2.62	2.51	2.18	1.98	1.73
25	4.24	3.38	2.99	2.76	2.60	2.49	2.16	1.96	1.71
26	4.22	3.37	2.98	2.74	2.59	2.47	2.15	1.95	1.69
27	4.21	3.35	2.96	2.73	2.57	2.46	2.13	1.93	1.67
28	4.20	3.34	2.95	2.71	2.56	2.44	2.12	1.91	1.65
29	4.18	3.33	2.93	2.70	2.54	2.43	2.10	1.90	1.64
30	4.17	3.32	2.92	2.69	2.53	2.42	2.09	1.89	1.62
40	4.03	3.23	2.84	2.61	2.45	2.34	2.00	1.79	1.51

Variance ratios for $p = 0.05$ (cont.)

n_2	n_1								
	1	2	3	4	5	6	12	24	∞
60	4.00	3.15	2.76	2.52	2.37	2.25	1.92	1.70	1.39
120	3.92	3.07	2.68	2.45	2.29	2.17	1.83	1.61	1.25
∞	3.84	2.99	2.60	2.37	2.21	2.10	1.75	1.52	1.00

Variance ratios for $p = 0.01$

n_2	n_1								
	1	2	3	4	5	6	12	24	∞
1	4052	4999	5403	5626	5764	5859	6106	6234	6366
2	98.50	99.00	99.17	99.25	99.30	99.33	99.42	99.46	99.50
3	34.12	30.82	29.46	28.71	28.24	27.91	27.05	26.60	26.12
4	21.20	18.00	16.69	15.98	15.52	15.21	14.37	13.93	13.46
5	16.26	13.27	12.06	11.39	10.97	10.67	9.89	9.47	9.02
6	13.74	10.92	9.78	9.15	8.75	8.47	7.72	7.31	6.88
7	12.25	9.55	8.45	7.85	7.46	7.19	6.47	6.07	5.65
8	11.26	8.65	7.59	7.01	6.63	6.37	5.67	5.28	4.86
9	10.56	8.02	6.99	6.42	6.06	5.80	5.11	4.73	4.31
10	10.04	7.56	6.55	5.99	5.64	5.39	4.71	4.33	3.91
11	9.65	7.20	6.22	5.67	5.32	5.07	4.40	4.02	3.60
12	9.33	6.93	5.95	5.41	5.06	4.82	4.16	3.78	3.36
13	9.07	6.70	5.74	5.20	4.86	4.62	3.96	3.59	3.16
14	8.86	6.51	5.56	5.03	4.69	4.46	3.80	3.43	3.00
15	8.68	6.36	5.42	4.89	4.56	4.32	3.67	3.29	2.87
16	8.53	6.23	5.29	4.77	4.44	4.20	3.55	3.18	2.75
17	8.40	6.11	5.18	4.67	4.34	4.10	3.45	3.08	2.65
18	8.28	6.01	5.09	4.58	4.25	4.01	3.37	3.00	2.57
19	8.18	5.93	5.01	4.50	4.17	3.94	3.30	2.92	2.49
20	8.10	5.85	4.94	4.43	4.10	3.87	3.23	2.86	2.42
21	8.02	5.78	4.87	4.37	4.04	3.81	3.17	2.80	2.36
22	7.94	5.72	4.82	4.31	3.99	3.76	3.12	2.75	2.31
23	7.88	5.66	4.76	4.26	3.94	3.71	3.07	2.70	2.26
24	7.82	5.61	4.72	4.22	3.90	3.67	3.03	2.66	2.21
25	7.77	5.57	4.68	4.18	3.86	3.63	2.99	2.62	2.17
26	7.72	5.53	4.64	4.14	3.82	3.59	2.96	2.58	2.13
27	7.68	5.49	4.60	4.11	3.78	3.56	2.93	2.55	2.10
28	7.64	5.45	4.57	4.07	3.75	3.53	2.90	2.52	2.06
29	7.60	5.42	4.54	4.04	3.73	3.50	2.87	2.49	2.03
30	7.56	5.39	4.51	4.02	3.70	3.47	2.84	2.47	2.01
40	7.31	5.18	4.31	3.83	3.51	3.29	2.66	2.29	1.80
60	7.08	4.98	4.13	3.65	3.34	3.12	2.50	2.12	1.60
120	6.85	4.79	3.95	3.48	3.17	2.96	2.34	1.95	1.38
∞	6.64	4.60	3.78	3.32	3.02	2.80	2.18	1.79	1.00

Variance ratios for $p = 0.001$

n_2	n_1								
	1	2	3	4	5	6	12	24	∞
1	4×10^5	5×10^5	5×10^5	5×10^5	5×10^5	5×10^5	6×10^5	6×10^5	6×10^5
2	998.5	999.0	999.2	999.2	999.3	999.3	999.4	999.5	999.5
3	167.0	148.5	141.1	137.1	134.6	132.8	128.3	125.9	123.5
4	74.14	61.25	56.18	53.44	51.71	50.53	47.41	45.77	44.05
5	47.18	37.12	33.20	31.09	29.75	28.84	26.42	25.14	23.78
6	35.51	27.00	23.70	21.92	20.81	20.03	17.99	16.89	15.75
7	29.25	21.69	18.77	17.19	16.21	15.52	13.71	12.73	11.60
8	25.42	18.49	15.83	14.39	13.49	12.86	11.19	10.30	9.34
9	22.86	16.39	13.90	12.56	11.71	11.13	9.57	8.72	7.81
10	21.04	14.91	12.55	11.28	10.48	9.92	8.45	7.64	6.76
11	19.69	13.81	11.56	10.35	9.58	9.05	7.63	6.85	6.00
12	18.64	12.97	10.80	9.63	8.89	8.38	7.00	6.25	5.42
13	17.81	12.31	10.21	9.07	8.35	7.86	6.52	5.78	4.97
14	17.14	11.78	7.73	8.62	7.92	7.43	6.13	5.41	4.60
15	16.59	11.34	9.34	8.25	7.57	7.09	5.81	5.10	4.31
16	16.12	10.97	9.00	7.94	7.27	6.81	5.55	4.85	4.06
17	15.72	10.66	8.73	7.68	7.02	6.56	5.32	4.63	3.85
18	15.38	10.39	8.49	7.46	6.81	6.35	5.13	4.45	3.67
19	15.08	10.16	8.28	7.26	6.62	6.18	4.97	4.29	3.52
20	14.82	9.95	8.10	7.10	6.46	6.02	4.82	4.15	3.38
21	14.59	9.77	7.94	6.95	6.32	5.88	4.70	4.03	3.26
22	14.38	9.61	7.80	6.81	6.19	5.76	4.58	3.92	3.15
23	14.19	9.47	7.67	6.69	6.08	5.65	4.48	3.82	3.05
24	14.03	9.34	7.55	6.59	5.98	5.55	4.39	3.74	2.97
25	13.88	9.22	7.45	6.49	5.88	5.46	4.31	3.66	2.89
26	13.74	9.12	7.36	6.41	5.80	5.38	4.24	3.59	2.82
27	13.61	9.02	7.27	6.33	5.73	5.31	4.17	3.52	2.75
28	13.50	8.93	7.19	6.25	5.66	5.24	4.11	3.46	2.70
29	13.39	8.85	7.12	6.19	5.59	5.18	4.05	3.41	2.64
30	13.29	8.77	7.05	6.12	5.53	5.12	4.00	3.36	2.59
40	12.61	8.25	6.60	5.70	5.13	4.73	3.64	3.01	2.23
60	11.97	7.76	6.17	5.31	4.76	4.37	3.31	2.69	1.90
120	11.38	7.32	5.79	4.95	4.42	4.04	3.02	2.40	1.54
∞	10.83	6.91	5.42	4.62	4.10	3.74	2.74	2.13	1.00

Appendix 4
Mann–Whitney U tables

These tables should be used if both n_1 and n_2 are less than 20. If n_1 and n_2 exceed 20 then use the z-statistic and the table in Appendix 2. The probability associated with given U values are tabulated for n_1 in each n_2 table.

$n_2 = 3$

U	n_1 1	2	3
0	.250	.100	.050
1	.500	.200	.100
2	.750	.400	.200
3		.600	.350
4			.500
5			.650

$n_2 = 4$

U	n_1 1	2	3	4
0	.200	.067	.028	.014
1	.400	.133	.057	.029
2	.600	.267	.144	.057
3		.400	.200	.100
4		.600	.314	.171
5			.429	.243
6			.571	.343
7				.443
8				.557

$n_2 = 5$

	n_1				
U	1	2	3	4	5
0	.167	.047	.018	.008	.004
1	.323	.095	.036	.016	.008
2	.500	.190	.071	.032	.016
3	.667	.286	.125	.056	.023
4		.429	.196	.095	.048
5		.571	.286	.143	.075
6			.393	.206	.111
7			.500	.278	.155
8			.607	.365	.210
9				.452	.271
10				.548	.345
11					.421
12					.500
13					.579

$n_2 = 6$

	n_1					
U	1	2	3	4	5	6
0	.143	.036	.012	.005	.002	.001
1	.286	.071	.024	.010	.004	.002
2	.428	.143	.048	.019	.009	.004
3	.571	.214	.083	.033	.015	.008
4		.321	.131	.057	.026	.013
5		.429	.190	.086	.041	.021
6		.571	.274	.129	.063	.032
7			.357	.176	.089	.017
8			.452	.238	.123	.066
9			.548	.305	.165	.090
10				.381	.214	.120
11				.457	.268	.155
12				.545	.331	.197
13					.396	.242
14					.465	.291
15					.535	.350
16						.409
17						.469
18						.531

$n_2 = 7$

U	n_1						
	1	2	3	4	5	6	7
0	.125	.028	.008	.003	.001	.001	.000
1	.250	.056	.017	.006	.003	.001	.001
2	.375	.111	.033	.012	.005	.002	.001
3	.500	.167	.058	.021	.009	.004	.002
4	.625	.250	.092	.036	.015	.007	.003
5		.333	.133	.055	.024	.011	.006
6		.444	.192	.082	.037	.017	.009
7		.556	.258	.115	.053	.026	.013
8			.333	.158	.074	.037	.019
9			.417	.206	.101	.051	.027
10			.500	.264	.143	.069	.036
11			.583	.324	.172	.090	.049
12				.394	.216	.117	.064
13				.464	.256	.147	.082
14				.538	.319	.183	.104
15					.378	.223	.130
16					.438	.267	.159
17					.500	.314	.191
18					.562	.365	.228
19						.418	.267
20						.473	.310
21						.527	.355
22							.402
23							.452
24							.500
25							.549

$n_2 = 8$

U	n_1							
	1	2	3	4	5	6	7	8
0	.111	.022	.006	.001	.000	.000	.000	.000
1	.222	.044	.012	.004	.002	.001	.000	.000
2	.333	.089	.024	.008	.003	.001	.001	.000
3	.444	.133	.042	.014	.005	.002	.001	.001
4	.556	.200	.067	.024	.009	.004	.002	.001
5		.267	.097	.036	.015	.006	.003	.001
6		.356	.139	.055	.023	.010	.005	.002
7		.444	.188	.077	.033	.015	.007	.003
8		.556	.248	.107	.047	.021	.010	.005

$n_2 = 8$ (cont.)

U	n_1 1	2	3	4	5	6	7	8
9			.315	.141	.064	.030	.014	.007
10			.387	.184	.085	.041	.020	.010
11			.461	.230	.111	.054	.027	.014
12			.539	.285	.142	.071	.036	.019
13				.341	.177	.091	.047	.025
14				.404	.217	.114	.060	.032
15				.467	.262	.141	.076	.041
16				.533	.311	.172	.095	.052
17					.362	.207	.116	.065
18					.416	.245	.140	.080
19					.472	.286	.168	.097
20					.528	.331	.198	.117
21						.377	.232	.139
22						.426	.268	.164
23						.475	.306	.191
24						.525	.347	.221
25							.389	.253
26							.433	.287
27							.478	.323
28							.522	.360
29								.399
30								.439
31								.480
32								.520

The following three tables show critical U values for $n_2 = 9$ to 20 tabulated against $n_1 = 1$ to 20.

$p = 0.05$ for a one-tail test and 0.1 for a two-tail test

n_1	n_2 9	10	11	12	13	14	15	16	17	18	19	20
1	0	0	0	0	0	0	0	0	0	0	0	0
2	1	1	1	2	2	2	3	3	3	4	4	4
3	3	4	5	5	6	7	7	8	9	9	10	11
4	6	7	8	9	10	11	12	14	15	16	17	18
5	9	11	12	13	15	16	18	19	20	22	23	25
6	12	14	16	17	19	21	23	25	26	28	30	32
7	15	17	19	21	24	26	28	30	33	35	37	39

p = 0.05 *for a one-tail test and* 0.1 *for a two-tail test (cont.)*

n_1	n_2 9	10	11	12	13	14	15	16	17	18	19	20
8	18	20	23	26	28	31	33	36	39	41	44	47
9	21	24	27	30	33	36	39	42	45	48	51	54
10	24	27	31	34	37	41	44	48	51	55	58	62
11	27	31	34	38	42	46	50	54	57	61	63	69
12	30	34	38	42	47	51	55	60	64	68	72	77
13	33	37	42	47	51	56	61	65	70	75	80	84
14	36	41	46	51	56	61	66	71	77	82	87	92
15	39	44	50	55	61	66	72	77	83	88	94	100
16	42	48	54	60	65	71	77	83	89	94	101	107
17	45	51	57	64	70	77	83	89	96	102	109	115
18	48	55	61	68	75	82	88	95	102	109	116	123
19	51	58	63	72	80	87	94	101	109	116	123	130
20	54	62	69	77	84	92	100	107	115	123	130	138

p = 0.01 *for a one-tail test and* 0.02 *for a two-tail test*

n_1	n_2 9	10	11	12	13	14	15	16	17	18	19	20
1	0	0	0	0	0	0	0	0	0	0	0	0
2	0	0	0	0	0	0	0	0	0	0	1	1
3	1	1	1	2	2	2	2	3	4	4	4	5
4	3	3	4	5	5	6	7	7	8	9	9	10
5	5	6	7	8	9	10	11	12	13	14	15	16
6	7	8	9	11	12	13	15	16	18	19	20	22
7	9	11	12	14	16	18	19	21	23	24	26	18
8	11	13	15	17	20	22	24	26	28	30	32	34
9	14	16	18	21	23	26	28	31	33	36	38	40
10	16	19	22	24	27	30	33	36	38	41	44	47
11	18	22	25	28	31	34	37	41	44	47	50	53
12	21	24	28	31	35	38	42	46	49	53	56	60
13	23	27	31	35	39	43	47	51	55	59	63	67
14	26	30	34	38	43	47	51	56	60	65	69	73
15	28	33	37	42	47	51	56	61	66	70	75	80
16	31	36	41	46	51	56	61	66	71	76	82	87
17	33	38	44	49	55	60	66	71	77	82	88	93
18	36	41	47	53	59	65	70	76	82	88	94	100
19	38	44	50	56	63	69	75	82	88	94	101	107
20	40	47	53	60	67	73	80	87	93	100	107	114

$p = 0.001$ *for a one-tail test and* 0.002 *for a two-tail test*

n_1	n_2											
	9	10	11	12	13	14	15	16	17	18	19	20
1	0	0	0	0	0	0	0	0	0	0	0	0
2	0	0	0	0	0	0	0	0	0	0	0	0
3	0	0	0	0	0	0	0	0	0	0	0	0
4	0	0	0	0	1	1	1	2	2	3	3	3
5	1	1	2	2	3	3	4	5	5	6	7	7
6	2	3	4	4	5	6	7	8	9	10	11	12
7	3	5	6	7	8	9	10	11	13	14	15	16
8	5	6	8	9	11	12	14	15	17	18	20	21
9	7	8	10	12	14	15	17	19	21	23	25	26
10	8	10	12	14	17	19	21	23	25	27	29	32
11	10	12	15	17	20	22	24	27	29	32	34	37
12	12	14	17	20	23	25	28	31	34	37	40	42
13	14	17	20	23	26	29	32	35	38	42	45	48
14	15	19	22	25	29	32	36	39	43	46	50	54
15	17	21	24	28	32	36	40	43	47	51	55	59
16	19	23	27	31	35	39	43	48	52	56	60	65
17	21	25	29	34	38	43	47	52	57	61	66	70
18	23	27	32	37	42	46	51	56	61	66	71	76
19	25	29	34	40	45	50	55	60	66	71	77	82
20	26	32	37	42	48	54	59	65	70	76	82	88

Appendix 5

Asymptotic Kolmogorov–Smirnov table

The following table shows critical difference values D_{crit} with their associated probabilities p.

D_{crit}	p
1.0727	0.200
1.2238	0.100
1.3581	0.050
1.5174	0.020
1.6276	0.010
1.7317	0.005
1.8585	0.002
1.9526	0.001

D_{crit} is calculated as D/s with the following formula:

$$D_{crit} = \frac{D}{s} = \frac{\max|F_{n_1} - F_{n_2}|}{\sqrt{(n_1 + n_2)/n_1 n_2}},$$

where D is the observed maximum difference between two normalised cumulative frequency distributions and s is the standard deviation. The number of observations must exceed 25 in each distribution if this formula is used. If either is less than 25, use the two-tail Kolmogorov–Smirnov tables.

Two-tail Kolmogorov–Smirnov tables

In Table 42, n_1 (columns) and n_2 (rows) are the numbers in the larger and smaller cumulative frequency distributions (respectively) below which the difference between the two is not significant at the 5% probability level. For example, if $n_1 = 18$ and $n_2 = 6$ then the critical distance is 0.61; if the maximum observed difference between the cumulative frequency distributions (both normalised to unity) is less than 0.61 then there is no significant difference at $p = 0.05$.

Table 42. D_{crit} values at the 5% probability level

n_2 \ n_1	2	3	4	5	6	7	8	9	10	11	12	13
2	−	−	1.00	1.00	1.00	1.00	1.00	1.00	1.00	1.00	1.00	1.00
3	−	−	−	1.00	1.00	1.00	0.950	0.950	0.950	0.900	0.900	0.900
4	−	−	−	1.00	0.900	0.900	0.875	0.833	0.750	0.750	0.750	0.750
5	−	−	−	1.00	0.833	0.800	0.800	0.777	0.750	0.750	0.717	0.692
6	−	−	−	−	0.833	0.800	0.777	0.722	0.700	0.667	0.667	0.667
7	−	−	−	−	−	0.833	0.750	0.700	0.667	0.667	0.631	0.615
8	−	−	−	−	−	−	0.750	0.667	0.667	0.630	0.625	0.596
9	−	−	−	−	−	−	−	0.660	0.650	0.623	0.583	0.565
10	−	−	−	−	−	−	−	−	0.630	0.615	0.550	0.550
11	−	−	−	−	−	−	−	−	−	0.585	0.545	0.535
12	−	−	−	−	−	−	−	−	−	−	0.540	0.530
13	−	−	−	−	−	−	−	−	−	−	−	0.530

n_2 \ n_1	14	15	16	17	18	19	20	21	22	23	24	25
2	0.940	0.940	0.940	0.940	0.940	0.940	0.940	0.930	0.930	0.920	0.920	0.920
3	0.860	0.850	0.830	0.825	0.820	0.820	0.810	0.810	0.800	0.800	0.800	0.800
4	0.750	0.733	0.720	0.710	0.700	0.700	0.700	0.700	0.700	0.700	0.690	0.680
5	0.667	0.667	0.667	0.660	0.660	0.650	0.650	0.650	0.640	0.640	0.640	0.640
6	0.643	0.633	0.625	0.610	0.610	0.610	0.600	0.595	0.591	0.590	0.590	0.590
7	0.643	0.600	0.571	0.571	0.571	0.571	0.564	0.551	0.550	0.550	0.547	0.545
8	0.571	0.558	0.547	0.566	0.555	0.540	0.540	0.535	0.534	0.532	0.530	0.520
9	0.555	0.555	0.541	0.536	0.529	0.526	0.517	0.523	0.510	0.510	0.510	0.507
10	0.532	0.525	0.525	0.523	0.511	0.508	0.505	0.500	0.500	0.496	0.491	0.490
11	0.530	0.517	0.506	0.500	0.500	0.488	0.486	0.485	0.481	0.478	0.470	0.469
12	0.520	0.510	0.500	0.500	0.500	0.483	0.476	0.474	0.470	0.465	0.463	0.446
13	0.570	0.510	0.486	0.475	0.470	0.461	0.461	0.461	0.455	0.451	0.448	0.428
14	0.565	0.500	0.480	0.466	0.460	0.455	0.450	0.448	0.442	0.440	0.435	0.427
15	−	0.480	0.475	0.460	0.455	0.450	0.450	0.438	0.436	0.433	0.433	0.427
16	−	−	0.465	0.456	0.444	0.437	0.437	0.432	0.426	0.426	0.420	0.417
17	−	−	−	0.450	0.456	0.440	0.436	0.429	0.422	0.420	0.414	0.410
18	−	−	−	−	0.430	0.430	0.422	0.420	0.414	0.410	0.410	0.400
19	−	−	−	−	−	0.425	0.421	0.415	0.404	0.404	0.401	0.400
20	−	−	−	−	−	−	0.420	0.410	0.400	0.400	0.400	0.400
21	−	−	−	−	−	−	−	0.410	0.400	0.391	0.390	0.385
22	−	−	−	−	−	−	−	−	0.395	0.383	0.380	0.380
23	−	−	−	−	−	−	−	−	−	0.380	0.375	0.375
24	−	−	−	−	−	−	−	−	−	−	0.370	0.375
25	−	−	−	−	−	−	−	−	−	−	−	0.370

In Table 43, n_1 (columns) and n_2 (rows) are the numbers in the larger and smaller cumulative frequency distributions (respectively) below which the difference between the two is not significant at the 1% probability level. For example, if $n_1 = 23$ and $n_2 = 15$ then the critical distance is 0.52; if the maximum observed difference between the cumulative frequency distributions (both normalised to unity) is less than 0.52 then there is no significant difference at $p = 0.01$.

Table 43. D_{crit} values at the 1% probability level

n_2	n_1 2	3	4	5	6	7	8	9	10	11	12	13
2	–	–	1.00	1.00	1.00	1.00	1.00	1.00	1.00	1.00	1.00	1.00
3	–	–	–	1.00	1.00	1.00	1.00	1.00	1.00	1.00	1.00	1.00
4	–	–	–	1.00	1.00	1.00	1.00	1.00	0.920	0.920	0.920	0.920
5	–	–	–	1.00	1.00	1.00	0.888	0.888	0.888	0.833	0.833	0.800
6	–	–	–	–	1.00	0.875	0.875	0.833	0.833	0.833	0.833	0.769
7	–	–	–	–	–	0.875	0.875	0.777	0.766	0.757	0.740	0.730
8	–	–	–	–	–	–	0.875	0.777	0.750	0.727	0.708	0.692
9	–	–	–	–	–	–	–	0.777	0.700	0.707	0.694	0.666
10	–	–	–	–	–	–	–	–	0.700	0.700	0.666	0.646
11	–	–	–	–	–	–	–	–	–	0.700	0.666	0.636
12	–	–	–	–	–	–	–	–	–	–	0.666	0.630
13	–	–	–	–	–	–	–	–	–	–	–	0.600

n_2	n_1 14	15	16	17	18	19	20	21	22	23	24	25
2	1.00	1.00	1.00	1.00	1.00	1.00	1.00	1.00	1.00	1.00	1.00	1.00
3	1.00	0.950	0.950	0.950	0.950	0.950	0.950	0.910	0.910	0.910	0.910	0.910
4	0.920	0.882	0.882	0.882	0.857	0.857	0.857	0.857	0.840	0.840	0.840	0.840
5	0.800	0.800	0.800	0.800	0.780	0.777	0.760	0.760	0.756	0.756	0.750	0.750
6	0.766	0.766	0.750	0.736	0.733	0.733	0.733	0.724	0.714	0.713	0.713	0.713
7	0.720	0.714	0.706	0.690	0.684	0.666	0.666	0.666	0.666	0.666	0.666	0.657
8	0.678	0.675	0.675	0.657	0.652	0.650	0.650	0.637	0.636	0.625	0.625	0.625
9	0.666	0.666	0.653	0.647	0.636	0.626	0.616	0.616	0.616	0.611	0.611	0.600
10	0.643	0.630	0.625	0.623	0.600	0.594	0.594	0.594	0.594	0.591	0.583	0.583
11	0.623	0.618	0.602	0.590	0.590	0.584	0.580	0.580	0.570	0.561	0.560	0.560
12	0.619	0.600	0.600	0.583	0.583	0.570	0.570	0.560	0.560	0.555	0.550	0.550
13	0.600	0.589	0.581	0.574	0.559	0.558	0.550	0.549	0.545	0.538	0.532	0.529
14	0.590	0.585	0.562	0.562	0.556	0.556	0.542	0.540	0.532	0.528	0.523	0.520
15	–	0.590	0.554	0.550	0.544	0.533	0.533	0.533	0.524	0.520	0.520	0.520
16	–	–	0.550	0.536	0.535	0.526	0.525	0.515	0.511	0.508	0.505	0.497
17	–	–	–	0.530	0.525	0.514	0.514	0.504	0.500	0.500	0.497	0.487
18	–	–	–	–	0.520	0.514	0.505	0.500	0.495	0.493	0.490	0.480
19	–	–	–	–	–	0.510	0.492	0.488	0.485	0.478	0.478	0.470
20	–	–	–	–	–	–	0.490	0.485	0.482	0.476	0.475	0.470
21	–	–	–	–	–	–	–	0.485	0.481	0.476	0.475	0.470
22	–	–	–	–	–	–	–	–	0.470	0.470	0.458	0.454
23	–	–	–	–	–	–	–	–	–	0.465	0.451	0.450
24	–	–	–	–	–	–	–	–	–	–	0.450	0.430
25	–	–	–	–	–	–	–	–	–	–	–	0.442

Appendix 6
Regression analysis routine for *y* on *x*

```
C    ...Micro VAX-3 F77 version 09-DEC-89  ************************************
C    ...Modified from PDP FORTRAN IV version 01-MAR-76
     SUBROUTINE REGRES(NPOINT,X,Y,XBAR,YBAR,SLOPE,YINT,C_COEF
     1,SIGMAX,SIGMAY,SEY)
C  Inputs to routine are  NPOINT  = Number of points
C                         X       = array of NPOINT x-values
C                         Y       = array of NPOINT y-values
C  Outputs from routine   XBAR    = mean of the x-values
C                         YBAR    = mean of the y-values
C                         SLOPE   = SLOPE
C                         YINT    = Y-axis intercept
C                         C_COEF  = correlation coefficient, r
C                         SIGMAX  = standard deviation of x-values
C                         SIGMAY  = standard deviation of y-values
C                         SEY     = standard error of Y
     DIMENSION X(NPOINT),Y(NPOINT)
     SF1(A,B,C)=(A*B)-C
     SF2(A,B)=A+(B*B)
     A=1./FLOAT(NPOINT)
     XBAR=0.
     SUMXSQ=0.
     YBAR=0.
     SUMYSQ=0.
     SUMXY=0.
        DO I=1,NPOINT
           XBAR=XBAR+X(I)
           SUMXSQ=SF2(SUMXSQ,X(I))
           YBAR=YBAR+Y(I)
           SUMYSQ=SF2(SUMYSQ,Y(I))
           SUMXY=SUMXY+(X(I)*Y(I))
        END DO
     XBAR=XBAR*A
     SIGMAX=SQRT(SF1(A,SUMXSQ,XBAR*XBAR))
     YBAR=YBAR*A
     SIGMAY=SQRT(SF1(A,SUMYSQ,YBAR*YBAR))
     C_COEF=SF1(A,SUMXY,XBAR*YBAR)/(SIGMAX*SIGMAY)
     SLOPE=(C_COEF*SIGMAY)/SIGMAX
     YINT=YBAR-(SLOPE*XBAR)
     SEY=SIGMAY*SQRT((1.0-(C_COEF**2)))
     RETURN
     END
```

Appendix 7

Consider a data set that contains I x-values and an array of J y-values per x-value. Point P_{ij} in the data set can be specified by the coordinates (x_i, y_{ij}) with a frequency defined as f_{ij}. The total number of y-values in the ith array, n_i, is given by

$$n_i = \sum_1^J f_{ij}. \tag{A7.1}$$

The mean value of the y's in the ith array, $\bar{y}_i$, is found by summing the array and dividing by the total frequency:

$$n_i \bar{y}_i = \sum_1^J f_{ij} y_{ij}. \tag{A7.2}$$

The total sum of squares, $N\sigma_y^2$, for the deviations of the y-values from the mean of all the y-values, $\bar{y}$, is given by

$$N\sigma_y^2 = \sum_1^I \sum_1^J f_{ij}(y_{ij} - \bar{y})^2, \tag{A7.3}$$

where N is the total number of data points and σ_y^2 is the variance of all the y-values.

The total sum of squares given by the right-hand side (RHS) of equation (A7.3) can be partitioned into the sums of squares within and between arrays, where

$$N\sigma_y^2 = \sum_1^I \sum_1^J f_{ij}((y_{ij} - \bar{y}_i) + (\bar{y}_i - \bar{y}))^2, \tag{A7.4}$$

where $(y_{ij} - \bar{y}_i)$ represents the deviations of the J y-values from the mean, $\bar{y}_i$, in the ith array. Expanding the RHS of equation (A7.4), we have

$$\sum_1^I \sum_1^J f_{ij}(y_{ij} - \bar{y}_i)^2 - 2\left(\sum_1^J f_{ij}(y_{ij} - \bar{y}_i) \sum_1^I f_{ij}(\bar{y}_i - \bar{y})\right) + \sum_1^I \sum_1^J f_{ij}(\bar{y}_i - \bar{y})^2.$$

Referring to equations (A7.1) and (A7.2), it can be seen that the term

$$\sum_1^J f_{ij}(y_{ij} - \bar{y}_i) = 0,$$

as $\sum_1^J f_{ij} = n_i$ and $\sum_1^J f_{ij} y_{ij} = n_i \bar{y}_i$. Thus, the product term vanishes and

$$N\sigma_y^2 = \sum_1^I \sum_1^J f_{ij}(y_{ij} - \bar{y}_i)^2 + \sum_1^I n_i(\bar{y}_i - \bar{y})^2. \tag{A7.5}$$

The first term on the RHS of equation (A7.5) is the sum of squares within arrays, and the second term can also be resolved into two components. Let Y_i be the estimated value of y_i from a regression of y on x; then

$$\sum_1^I n_i(\bar{y}_i - \bar{y})^2 = \sum_1^I n_i((\bar{y}_i - Y_i) + (Y_i - \bar{y}))^2$$

$$= \sum_1^I n_i(Y_i - \bar{y})^2 + \sum_1^I n_i(\bar{y}_i - Y_i)^2. \tag{A7.6}$$

The product term in the expansion of this equation also vanishes for each array as in equation (A7.4). The first term on the RHS of equation (A7.6) represents the sum of squares due to linear regression, and the second term is the sum of squared deviations of the mean values of arrays from the regression line. We can now tabulate the sources of variation and their accompanying sums of squares from equations (A7.5) and (A7.6) as follows.

Source of variation	Sums of squares
Within arrays	$\sum_1^I \sum_1^J f_{ij}(y_{ij} - \bar{y}_i)^2$
Linear regression	$\sum_1^I n_i(Y_i - \bar{y})^2$
Deviation of means	$\sum_1^I n_i(\bar{y}_i - Y_i)^2$
Total	$\sum_1^I \sum_1^J f_{ij}(y_{ij} - \bar{y})^2$

Dividing these sums by the degrees of freedom, DF, associated with the respective sources of variation gives unbiased estimates of the variances. The total number of degrees of freedom is $N-1$. One degree of freedom is associated with the linear regression, which leaves $N-2$ DF for the within-arrays variation plus the deviations of means from the regression line. As there are I arrays of y's we have $N-I$ DF associated with the variation within arrays and hence $I-2$ DF associated with the deviations of means from the regression line. We thus obtain the following relationships:

$$\text{variance within arrays} = \frac{\sum_1^I \sum_1^J f_{ij}(y_{ij} - \bar{y}_i)^2}{N-I}, \tag{A7.7}$$

$$\text{variance of deviation of means} = \frac{\sum_1^I n_i(\bar{y}_i - Y_i)^2}{I-2}. \tag{A7.8}$$

If the regression is linear then the variance estimate obtained from equation (A7.8) will not differ significantly from that obtained using equation (A7.7). The variance ratio F used to test for this condition is given by

$$F = \frac{\sum_1^I n_i(\bar{y}_i - Y_i)^2}{\sum_1^I \sum_1^J f_{ij}(y_{ij} - \bar{y}_i)^2} \times \frac{N-I}{I-2}, \tag{A7.9}$$

with $I-2$ and $N-I$ degrees of freedom for the greater and lesser variance estimates, respectively. Equation (A7.9) can be simplified to

$$F = \frac{\eta^2 - r^2}{1 - \eta^2} \times \frac{N-I}{I-2},$$

where r is the correlation coefficient and η is the correlation ratio. The formula for r was given in Section 5.1.2, and that for η can be calculated as the ratio of the standard deviation of the weighted means of the arrays of y's, σ_w, to the standard deviation of all the y-values, σ_y. σ_w can be obtained from the last term of equation (A7.5), but it is more conveniently calculated as

$$\sigma_w = \sqrt{\left(\frac{1}{N} \sum_1^I \frac{A_i^2}{N_i}\right) - \bar{y}^2},$$

where A_i is the sum of the J y-values in the ith array of y's (see Section 5.1.5).

Appendix 8

Consider two distributions X and Y with six elements in each, where the individual frequencies are given by $x_1, x_2, x_3, x_4, x_5, x_6$ and $y_1, y_2, y_3, y_4, y_5, y_6$ respectively. Let there be totals of n_1 and n_2 cells in X and Y, respectively, and let the probability functions underlying each distribution be given by $f(x)_c$ and $f(y)_c$ where the subscript c denotes channel number. Also, let the two distributions overlap by three channels:

1	2	3	4	5	6	7	8	9
$f(x)_1 n_1$	$f(x)_2 n_1$	$f(x)_3 n_1$	$f(x)_4 n_1$	$f(x)_5 n_1$	$f(x)_6 n_1$			
			$f(y)_1 n_2$	$f(y)_2 n_2$	$f(y)_3 n_2$	$f(y)_4 n_2$	$f(y)_5 n_2$	$f(y)_6 n_2$

where the channel numbers are shown and where $f(x)_1 n_1 = x_1$, $f(y)_1 n_2 = y_1$, etc.

The cumulative frequency curve is obtained by successive addition, which is shown in the following table for the distribution of X and Y combined. The total frequency in each channel is obtained by summing the elements in each column.

1	2	3	4	5	6	7	8	9
$f(x)_1 n_1$	$f(x)_2 n_1$	$f(x)_3 n_1$	$f(x)_4 n_1$	$f(x)_5 n_1$	$f(x)_6 n_1$	$f(x)_6 n_1$	$f(x)_6 n_1$	$f(x)_6 n_1$
	$f(x)_1 n_1$	$f(x)_2 n_1$	$f(x)_3 n_1$	$f(x)_4 n_1$	$f(x)_5 n_1$	$f(x)_5 n_1$	$f(x)_5 n_1$	$f(x)_5 n_1$
		$f(x)_1 n_1$	$f(x)_2 n_1$	$f(x)_3 n_1$	$f(x)_4 n_1$	$f(x)_4 n_1$	$f(x)_4 n_1$	$f(x)_4 n_1$
			$f(x)_1 n_1$	$f(x)_2 n_1$	$f(x)_3 n_1$	$f(x)_3 n_1$	$f(x)_3 n_1$	$f(x)_3 n_1$
				$f(x)_1 n_1$	$f(x)_2 n_1$	$f(x)_2 n_1$	$f(x)_2 n_1$	$f(x)_2 n_1$
					$f(x)_1 n_1$	$f(x)_1 n_1$	$f(x)_1 n_1$	$f(x)_1 n_1$
			$f(y)_1 n_2$	$f(y)_2 n_2$	$f(y)_3 n_2$	$f(y)_4 n_2$	$f(y)_5 n_2$	$f(y)_6 n_2$
				$f(y)_1 n_2$	$f(y)_2 n_2$	$f(y)_3 n_2$	$f(y)_4 n_2$	$f(y)_5 n_2$
					$f(y)_1 n_2$	$f(y)_2 n_2$	$f(y)_3 n_2$	$f(y)_4 n_2$
						$f(y)_1 n_2$	$f(y)_2 n_2$	$f(y)_3 n_2$
							$f(y)_1 n_2$	$f(y)_2 n_2$
								$f(y)_1 n_2$

We can now see that the cumulative frequency in channel 1 is $f(x)_1 n_1 = x_1$ and that in channel 9 it is the total summation of both distributions, which is $n_1 + n_2$.

The cumulative frequency–difference distribution is obtained by subtracting the combined distribution $X+Y$ from the control X on a channel-by-channel basis. However, before we do this the latter must be normalised to the total number of cells in the combined distribution. This is effected by multiplying each element in the control distribution by the factor $(n_1+n_2)/n_1$. Thus, the cumulative frequency subtraction (CFS) in channel 1 is given by

$$\text{CFS}(1) = (((n_1+n_2)/n_1)f(x)_1 n_1) - f(x)_1 n_1,$$

which reduces to $\text{CFS}(1) = f(x)_1 n_2$. Similarly, for example, the cumulative frequency subtraction in channel 5, CFS(5), is given by

$$\text{CFS}(5) = n_2(f(x)_1 + f(x)_2 + f(x)_3 + f(x)_4 + f(x)_5) - n_2(f(y)_1 + f(y)_2).$$

The probability functions underlying both individual distributions have the same variance and are symmetrical about their means, so in this example

$$f(x)_1 = f(x)_6 = f(y)_1 = f(y)_6,$$

$$f(x)_2 = f(x)_5 = f(y)_2 = f(y)_5,$$

$$f(x)_3 = f(x)_4 = f(y)_3 = f(y)_4,$$

and making the relevant substitutions in CFS(5) we obtain

$$\text{CFS}(5) = n_2(f(x)_1 + f(x)_2 + f(x)_3 + f(x)_2) - n_2(f(x)_1 + f(x)_2)$$
$$= n_2(f(x_2 + 2f(x)_3).$$

We can now generate the whole cumulative frequency difference distribution:

$$\text{CFS}(0) = 0,$$

$$\text{CFS}(1) = (f(x)_1)n_2,$$

$$\text{CFS}(2) = (f(x)_1 + f(x)_2)n_2,$$

$$\text{CFS}(3) = (f(x)_1 + f(x)_2 + f(x)_3)n_2,$$

$$\text{CFS}(4) = \qquad (f(x)_2 + f(x)_3 + f(x)_3)n_2,$$

$$\text{CFS}(5) = \qquad (f(x)_2 + f(x)_3 + f(x)_3)n_2,$$

$$\text{CFS}(6) = (f(x)_1 + f(x)_2 + f(x)_3)n_2,$$

$$\text{CFS}(7) = (f(x)_1 + f(x)_2)n_2,$$

$$\text{CFS}(8) = (f(x)_1)n_2,$$

$$\text{CFS}(9) = 0.$$

It can be seen that the cumulative frequency–difference distribution is symmetrical and independent of n_1, the number of cells in the first distribution. Furthermore, no assumptions have been made with regard to the shape of the underlying probability functions $f(x)$ and $f(y)$. Only six intervals were used for each distribution in this illustration in order to maintain simplicity, but this choice also corresponds to three standard deviations about the mean of a Gaussian distribution. However, the distributions do not have to be Gaussian; they could be rectangular, triangular or any symmetrical shape that can be imagined. In our particular example we can see that the distribution has a mean D_m in channel 4.5. If we now refer back to the depiction of the overlapping distributions at the beginning of this appendix, we can see that the means of the X and Y distributions, $\bar{x}$ and $\bar{y}$, were in channels 3.5 and 6.5 respectively. If we now substitute the values of 4.5 and 3.5 for D_m and $\bar{x}$ (respectively) in equation (7.5) with a class interval of unity, we obtain

$$\bar{y} = 2(4.5 + 0.5) - 3.5 = 6.5$$

which, as we have seen, is the mean of the Y distribution.

Appendix 9

Two rate equations that require integration were encountered in Sections 10.3.2 and 10.5.1 and are reproduced here:

$$\frac{\delta[P]}{\delta t} + k_3[P] = \frac{k_2[E_0][S]}{[S] + K_m},$$

$$\frac{\delta S}{\delta t} + k_2 S = k_1 E.$$

These equations are in fact identical; they differ only in the numbers of the various constants and the "names" of those constants. Their integration is identical, and in order to treat them as one I'll make a number of changes of variable. Let [P] and S on the left-hand side (LHS) of both equations be represented by the symbol A. Thus $\delta[P]/\delta t$ and $\delta S/\delta t$ are replaced by $\delta A/\delta t$. Let the rate constants on the LHS of both equations, k_3 and k_2 respectively, be represented by α. Let the rate constants on the right-hand side (RHS) of both equations, k_3 and k_2 respectively, be represented by β. Finally, let $([E_0][S])/([S] + K_m)$ and E, top and bottom equations respectively, be represented by the symbol B. Both equations can now be written as

$$\frac{\delta A}{\delta t} + \alpha A = \beta B.$$

Unfortunately, this equation is not in an integrable form. The trick is to transform it so that it can be integrated, and to do this we must make a slight diversion. The Law of Leibniz, who did almost as much in this field as Newton, states that the rate of change of the product of two variables is equal to the product of the first variable and the rate of change of the second added to the product of the second variable and the rate of change of the first. This is another brainful, which is more easily appreciated in symbolic form as follows:

$$\frac{\delta(UV)}{\delta t} = U \times \frac{\delta V}{\delta t} + V \times \frac{\delta U}{\delta t}.$$

We transform the equation we want to integrate by multiplying each term by $e^{\alpha t}$ to give

$$(e^{\alpha t})\frac{\delta A}{\delta t} + \alpha(e^{\alpha t})A = \beta(e^{\alpha t})B.$$

You might well ask "How the hell does this help?" But, in fact, it is extremely clever (and due to Leibniz). A in our equation is equivalent to V in Leibniz's equation and $\delta A/\delta t$ is equivalent to $\delta V/\delta t$, which is not too difficult. The really crafty bit is that the term $e^{\alpha t}$ in our equation is equivalent to U which, when differentiated, gives $\alpha(e^{\alpha t})$ or the rate of change of U, $\delta U/\delta t$. Thus, our equation now reduces to

$$\frac{\delta[A(e^{\alpha t})]}{\delta t} = \beta B(e^{\alpha t}).$$

This can now be integrated, remembering to add the constant C, to yield

$$A(e^{\alpha t}) = \frac{\alpha}{\beta}B(e^{\alpha t}) + C,$$

and we now divide through by $e^{\alpha t}$ to obtain

$$A = \frac{\alpha}{\beta}B + Ce^{-\alpha t}.$$

At time $t = 0.0$ the expression $e^{-\alpha t}$ is unity and there will be no drug or substrate in the cells. Therefore $A = 0.0$ and hence

$$C = -\frac{\alpha}{\beta}B.$$

Substituting this solution for C in the previous equation gives

$$A = \left[\frac{\alpha}{\beta}E\right] - \left[\frac{\alpha}{\beta}E\right]e^{-\alpha t}.$$

On simplification this now reduces to the familiar inverted exponential

$$A = \frac{\alpha}{\beta}B(1 - e^{-\alpha t}).$$

If we now make the relevant re-substitutions for the first rate equation, and change the symbol P to F to represent fluorescence, we have

$$F(t) = \frac{k_3[E_0] \times [S]}{k_2([S] + K_m)} \times (1 - e^{-k_3 t}).$$

Now, at infinite substrate concentration, $[S_\infty]$, the ratio $[S]/([S] + K_m)$ tends to unity and at infinite time the expression $e^{-k_3 t}$ tends to zero. Hence $[F_\infty]$, the asymptotic value for $F(t)$ at $t = \infty$ for $[S_\infty]$, will be given by

$$F_\infty = \frac{k_3}{k_2} E_0.$$

If we now substitute this into the previous equation we obtain

$$F(t) = F_\infty \left[\frac{[S]}{[S] + K_m} \right] \times (1 - e^{-k_3 t}).$$

Similarly, when we make the relevant re-substitutions in the second rate equation we get

$$S = \frac{k_1}{k_2} E (1 - e^{-k_2 t}),$$

where k_2 is the rate constant for efflux and where the uptake curve should reach an asymptote, Y_0, of $E(k_1/k_2)$.

References

Adler, A. J. and Kistiakowsky, G. B. (1961) Isolation and properties of pig liver ester-ases. *J. Biol. Chem.* 236: 3240-45.

Alt, F. W., Kellems, R. E. and Schimke, R. T. (1976) Synthesis and degradation of folate reductase in sensitive and methotrexate resistant lines of S-180 cells. *J. Biol. Chem.* 251: 3063-74.

Angerer, L. M. and Moudrianakis, E. N. (1972) Intercalation of ethidium bromide with whole and selectively deprotinized deoxynucleoproteins from calf thymus. *J. Mol. Biol.* 63: 505-21.

Assaraf, Y. G. and Schimke, R. T. (1987) Identification of methotrexate transport de-ficiency in mammalian cells using fluoresceinated methotrexate and flow cytometry. *Proc. Nat. Acad. Sci. USA* 84: 7154-58.

Assaraf, Y. G., Seamer, L. C. and Schimke, R. T. (1989) Characterization by flow cy-tometry of fluorescein-methotrexate transport in chinese hampster ovary cells. *Cy-tometry* 10: 50-55.

Babson, J. R. and Reed, D. J. (1978) Inactivation of glutathione reductase by 2-chloro-ethyl nitrosourea-derived isocyanates. *Biochem. Biophys. Res. Comm.* 83: 754-62.

Baisch, H. and Beck, H.-P. (1978) Comparison of cell kinetic parameters obtained by flow cytometry and autoradiography. In A. J. Valleron and P. D. M. MacDonald (eds.), *Biomathematics and Cell Kinetics*, p. 411. Amsterdam: Elsevier.

Baisch, H., Beck, H.-P., Christensen, I. J., Hartmann, N. R., Fried, J., Dean, P. N., Gray, J. W., Jett, J. H., Johnston, D. A., White, R. A., Nicolini, C., Zeitz, S. and Watson, J. V. (1982) A comparison of mathematical methods for the analysis of DNA histograms obtained by flow cytometry. *Cell and Tissue Kinet.* 15: 235-49.

Baisch, H., Gohde, W. and Linden, W. A. (1975) Analysis of PCP-data to determine the fraction of cells in the various phases of the cell cycle. *Rad. Environ. Biophys.* 12: 31-39.

Beck, H.-P. (1978) A new analytical method for determining duration of phases, rate of DNA-synthesis and degree of synchronization from flow cytometric data on syn-chronized cell populations. *Cell and Tissue Kinet.* 11: 139-48.

Beck, H.-P. (1981) Proliferation kinetics of perturbed cell populations determined by the bromodeoxyuridine-33258 technique: Radiotoxic effects of incorporated ^{3}H-thymi-dine. *Cytometry* 2: 170-74.

Begg, A. C., McNally, N. J., Schrieve, D. C. and Karcher, H. (1985) A method to mea-sure the duration of DNA synthesis and the potential doubling time from a single sample. *Cytometry* 6: 620-26.

Begleiter, A., Lam, H. Y. P. and Goldenberg, G. J. (1977) Mechanism of uptake of nitrosoureas by L5178Y lymphoblasts *in vitro. Cancer Res.* 37: 1022-27.

Beisker, W., Dolbeare, F. and Gray, J. W. (1987) An improved immunocytochemical procedure for high-sensitivity detection of incorporated bromodeoxyuridine. *Cytometry* 8: 235-39.

Berridge, M. J. and Irvine, R. F. (1984) Inositole triphosphate, a novel second messenger in intracellular signal transduction. *Nature* 312: 315-21.

Bohmer, R. M. (1979) Flow cytometric cell cycle analysis using the quenching of 33258 Hoechst fluorescence by bromodeoxyuridine incorporation. *Cell and Tissue Kinet.* 12: 101-10.

Brodie, S., Giron, J. and Latt, S. A. (1975) Estimation of accessibility of DNA in chromatin from fluorescence measurements of electronic excitation energy transfer. *Nature* 253: 470-71.

Brown, W. E. and Wold F. (1973) Alkyl isocyanates as active-site-specific reagents for serine proteases. Identification of the active-site serine as the site of reaction. *Biochem.* 12: 835-40.

Chambers Twentieth Century Dictionary (1976). Edinburgh: W & R Chambers.

Christensen, I., Hartmann, N. R., Keiding, N., Larsen, J. K., Noer, H. and Vindelov, L. (1978) Statistical analysis of DNA distributions from cell populations with partial synchrony. In D. Lutz (ed.), *Third International Symposium on Pulse-Cytophotometry,* pp. 71-78. Ghent: European Press.

Chused, T. M., Wilson, A. H., Greenblatt, D., Ishida, Y., Edison, L. J., Tsien, R. Y. and Finelman, F. D. (1978) Flow cytometric analysis of murine splenic B lymphocyte cytosolic free calcium response to anti-IgM and anti-IgD. *Cytometry* 8: 396-404.

Chused, T. M., Wilson, A. H., Seligman, B. E. and Tsien, R. Y. (1986) Probes for use in the study of leukocyte physiology by flow cytometry. In D. Lansing-Taylor et al. (eds.), *Applications of Fluorescence in the Biomedical Sciences,* pp. 531-44. New York: Alan R. Liss.

Cumber, P. M., Jacobs, A., Hoy, T., Fisher, J., Whittaker, J. A. and Tsuro, T. (1990) Expression of the multiple drug resistance gene (mdr-1) and epitope masking in chronic lymphatic leukaemia. *Brit. J. Haematol.* 76: 226-30.

Darzynkiewicz, Z., Andreeff, M., Traganos, F., Sharpless, T. and Melamed, M. R. (1978) Discrimination between cycling and noncycling lymphocytes by BUdR-suppressed acridine orange fluorescence in a flow cytometric system. *Exptl. Cell Res.* 115: 31-35.

Darzynkiewicz, A., Traganos, F. and Melamed, M. (1980) New cell cycle compartments identified by multiparameter flow cytometry. *Cytometry* 1: 98-108.

Darzynkiewicz, Z., Traganos, F., Sharpless, T. and Melamed, M. R. (1975) Conformation of RNA *in situ* as studied by acridine orange staining and automated cytophotometry. *Exptl. Cell Res.* 95: 143-53.

Darzynkiewicz, Z., Traganos, F., Sharpless, T. and Melamed, M. R. (1977a) Cell cycle related changes in nuclear chromatin of stimulated lymphocytes as measured by flow cytometry. *Cancer Research* 37: 4635-40.

Darzynkiewicz, Z., Traganos, F., Sharpless, T. and Melamed, M. R. (1977b) Different sensitivity of DNA *in situ* in interphase and metaphase chromatin to heat denaturation. *J. Cell Biol.* 73: 128-38.

Darzynkiewicz, Z., Traganos, F., Sharpless, T. and Melamed, M. R. (1977c) Recognition of cells in mitosis by flow cytometry. *J. Histochem. Cytochem.* 25: 875-80.

Dean, P. N., Dolbeare, F., Gratzner, H. G., Rice, G. C. and Gray, J. W. (1984) Cell cycle analysis using a monoclonal antibody to BrdUrd. *Cell and Tissue Kinet.* 17: 427-36.

Dean, P. N. and Jett, J. H. (1974) Mathematical analysis of DNA distributions derived from flow microfluorometry. *J. Cell Biol.* 60: 523-27.

Dean, P. N., Kolla, S. and van Dilla, M. A. (1989) Analysis of bivariate flow karyotypes. *Cytometry* 10: 109–23.

Dembo, M., Sirontak, F. M. and Moccia, D. M. (1984) Effects of metabolic deprivation on methotrexate transport in L1210 leukaemia cells: further evidence for separate influx and efflux systems and different energetic requirements. *J. Membrane Biol.* 78: 9–17.

de Moivre, A. (1756) In *Doctrines of Chance,* 3rd ed., p. 243. London: Millar. Royal Society Library, London.

Dive, C. (1988) Flow cytoenzymology with special reference to cancer chemotherapy. Ph.D. Thesis, Council for National Academic Awards, Gray's Inn Road, London.

Dive, C., Workman, P. and Watson, J. V. (1987) Novel dynamic flow cytoenzymological determination of intracellular esterase inhibition by BCNU and related isocyanates. *Biochem. Pharmacol.* 36: 3731–38.

Dive, C., Workman, P. and Watson, J. V. (1988) Inhibition of intracellular esterases by antitumour chloroethylnitrosoureas: measurement by flow cytometry and correlation with molecular carbamoylation activity. *Biochem. Pharmacol.* 37: 3987–93.

Dolbeare, F., Beisker, W., Pallavicini, M. G., Vanderlaan, M. and Gray, J. W. (1985) Ctyochemistry of bromodeoxyuridine/DNA analysis: stoichiometry and sensitivity. *Cytometry* 6: 521–30.

Dolbeare, F., Gratzner, H. G., Pallavicini, M. G. and Gray, J. W. (1983) Flow cytometric measurement of total DNA content and incorporated bromodeoxyuridine. *Proc. Nat. Acad. Sci. USA* 80: 5573–77.

Dolbeare, F. A. and Smith, R. E. (1971) Flow cytometric measurement of peptidases with use of 5-nitrosalicyladehyde and 4-methoxy-b-naphthylamine derivatives. *Clin. Chem.* 23: 1485–91.

Durand, R. and Olive, P. (1981) Flow cytometric studies of intracellular adriamycin in single cells *in vitro. Cancer Res.* 41: 3489–94.

Eadie, G. S. (1942) The inhibition of cholinesterase by physostigmine and prostigmine. *J. Biol. Chem.* 146: 85–93.

Eisenthal, R. and Cornish-Bowden, A. (1974) The direct linear plot. A new graphical procedure for estimating enzyme kinetic parameters. *Biochem. J.* 139: 715–20.

Fleichter, G. E. and Goerttler, K. (1986) Pitfalls in the preparation of nuclear suspension from paraffin-embedded tissue for flow cytometry. *Cytometry* 7: 616.

Fisher, G. A. (1961) Increased levels of folic acid reductase as a mechanism of resistance to amethopterin in leukaemic cells. *Biochem. Pharmacol.* 7: 75–80.

Fisher, G. A. (1962) Defective transport of amethopterin (methotrexate) as a mechanism of resistance to the antimetabolite in L5178Y leukaemia cells. *Biochem. Pharmacol.* 11: 1233–34.

Fisher, R. A. (1925a) Applications of "Student's" distribution. *Metron* 5: 90–104.

Fisher, R. A. (1925b) In *Statistical Methods for Research Workers,* 1st ed., Chapter V. Edinburgh: Oliver and Boyd.

Fisher, R. A. and Yates, F. (1963) *Statistical Tables for Biological, Agricultural and Medical Research,* 6th ed. Edinburgh: Oliver and Boyd.

Flintoff, W. E., Davidson, S. V. and Siminovitch, L. (1976) Isolation and partial characterization of three methotrexate-resistant phenotypes from chinese hamster ovary cells. *Somatic Cell Genet.* 2: 245–61.

Fox, M. H. (1980) A model for the computer analysis of synchronous DNA distributions obtained by flow cytometry. *Cytometry* 1: 71–77.

Fredericq, E. (1971) The chemical and physical properties of nucleohistones. In D. M. P. Phillips (ed.), *Histones,* p. 135. New York: Plenum Press.

Fried, J. (1976) Method for the quantitative evaluation of data from flow cytofluorometry. *Comp. Biomed. Res.* 9: 263-76.

Fried, J. (1977) Analysis of deoxyribonucleic acid histograms from flow cytometry. *J. Histochem. Cytochem.* 25: 942-51.

Fried, J. and Mandel, M. (1979) Multi-user system for analysis of data from flow cytometry. *Comput. Programs Biomed.* 10: 218-30.

Gapski, G. R., Whiteley, J. M., Rader, J. I., Cramer, P. L., Henderson, G. B., Neef, V. and Huennekens, F. M. (1975) Synthesis of a fluorescent derivative of amethopterin. *J. Med. Chem.* 18: 526-28.

Gaudray, P., Trotter, J. and Wahl, G. M. (1986) Fluorescent methotrexate labelling and flow cytometric analysis of cells containing low levels of dihydrofolate reductase. *J. Biol. Chem.* 261: 6285-92.

Gauss, K. F. (1809) *Theoria Motus Corporium Caelestium.* Hamburg: F. Perthes and I. H. Beesen. Cambridge University Library bench mark IV.1.16.

Gibson, N. W., Mattes, W. B. and Hartley, J. A. (1985) Identification of specific DNA lesions of chloroethylating agents: chloroethylnitrosoureas, chloroethylmethanesulphonates and chloroethylimidazotetrazines. *Pharma. Ther.* 31: 153-63.

Goldenberg, G. J. and Begleiter, A. (1984) Alterations of drug transport. In B. W. Fox and M. Fox (eds.), *Antitumour Drug Resistance,* pp. 241-98. Berlin: Springer.

Gratzner, H. G. (1982) Monoclonal antibody to 5-bromo- and 5-iododeoxyuridine: a new reagent for detecting DNA replication. *Science* 218: 474-75.

Gratzner, H. G. and Leif, R. C. (1981) An immunofluorescence method for monitoring DNA synthesis by flow cytometry. *Cytometry* 1: 385-89.

Gratzner, H. G., Leif, R. C., Ingram, D. J. and Castro, A. (1975) The use of antibody specific for bromodeoxyuridine for the immunofluorescent determination of DNA replication in single cells and chromosomes. *Exptl. Cell Res.* 95: 88-94.

Gray, J. W. (1974) Cell cycle analysis from computer synthesis of deoxyriboneucleic acid histograms. *J. Histochem. Cytochem.* 22: 642-50.

Gray, J. W. (1976) Cell-cycle analysis of perturbed cell populations; computer simulations of sequential DNA distribution. *Cell and Tissue Kinet.* 9: 499-510.

Gray, J. W. (1980) Flow cytometry and cell kinetics: relation to cancer therapy. In O. Laerum, T. Lindmo and E. Thorud (eds.), *Flow Cytometry IV,* p. 485. Bergen, Norway: Universitetforlaget.

Gray, J. W., Dean, P. N. and Mendelsohn, M. L. (1979) Quantitative cell-cycle analysis. In M. Melamed, P. Mullaney and M. L. Mendelsohn (eds.), *Flow Cytometry and Sorting,* p. 383. New York: Wiley.

Gray, J. W. and Mayall, B. H. (1985) *Monoclonal Antibodies Against Bromodeoxyuridine.* New York: Alan R. Liss.

Grynkiewicz, G., Poenie, M. and Tsien, R. Y. (1985) A new generation of Ca^{2+} indicators with greatly improved fluorescence properties. *J. Biol. Chem.* 260: 3440-50.

Guibault, G. G. and Kramer, D. N. (1966) Lipolysis of fluorescein and eosin esters. Kinetics of hydrolysis. *Analyt. Biochem.* 14: 28-32.

Haber, D. A., Beverly, S. M., Kiely, M. L. and Schimke, R. T. (1981) Properties of an altered dihydrofolate reductase encoded by amplified genes in cultured mouse fibroblasts. *J. Cell Biol.* 256: 9501-10.

Hamming, R. W. (1973) *Numerical Methods for Scientists and Engineers.* New York: McGraw-Hill.

Hanggi, U. J. and Littlefield, J. W. (1976) Altered gene regulation of the rate of synthesis of dihydrofolate reductase in methotrexate-resistant hamster cells. *J. Biol. Chem.* 251: 3075-82.

Hann, S. R., Thompson, C. B. and Eisenman, R. N. (1985) c-*myc* oncogene protein is independent of the cell cycle in human and avian cells. *Nature* 314: 366-69.

Hartmann, N. R. and Pederson, T. (1970) Analysis of the kinetics of granulosa cell populations in the mouse ovary. *Cell and Tissue Kinet.* 3: 1-11.

Hedley, D. W. (1989) Flow cytometry using paraffin-embedded tissue: five years on. *Cytometry* 10: 229-41.

Henderson, G. B., Russell, A. and Whiteley, J. M. (1980) A fluorescent derivative of methotrexate as an intracellular marker of dihydrofolate reductase in L1210 cells. *Arch. Biochem. Biophys.* 202: 29-34.

Herweijer, H., van den Engh, C. J. and Nooter, K. (1989) A rapid and sensitive flow cytometric method for the detection of multidrug-resistant cells. *Cytometry* 10: 463-68.

Hidderman, W., Schumann, J., Andreeff, M., Barlogie, B., Herman, C. J., Leif, R. C., Mayall, B. H., Murphy, R. F. and Sandberg, A. A. (1984) Convention on nomenclature for DNA cytometry. *Cytometry* 5: 445-46.

Hieber, L., Beck, H.-P. and Lucke-Huhle, C. (1981) G2-delay after irradiation with α-particles as studied in synchronized cultures and by the bromodeoxyuridine-33258H technique. *Cytometry* 2: 175-78.

Hill, H. D., Sumner, G. K. and Waters, M. D. (1968) An automated fluorimetric assay for alkaline phosphatase using 3-O-methyl fluorescein phosphate. *Analyt. Biochem.* 24: 9-14.

Hofstee, B. H. S. (1952) On the evaluation of the constants V_m and K_m in enzyme reactions. *Science* 116: 329-31.

Hopwood, D. (1985) Cell and tissue fixation, 1972-1982. *Histochem. J.* 17: 389-442.

Howard, A. and Pelc, S. R. (1951) Nuclear incorporation of ^{32}P as demonstrated by autoradiographs. *Exptl. Cell Res.* 2: 178-87.

Hulett, H. R., Bonner, W. A., Barrett, J. and Herzenberg, L. A. (1969) Cell sorting: automated separation of mammalian cells as a function of intracellular fluorescence. *Science* 166: 747-49.

Jett, J. H. (1978) Mathematical analysis of DNA histograms from asynchronous and synchronous cell populations. In D. Lutz (ed.), *Third International Symposium on Pulse-Cytophotometry*, p. 93. Ghent: European Press.

Johnston, D. A., White, R. A. and Barlogie, B. (1978) Automatic processing and interpretation of DNA distributions: comparison of several techniques. *Comp. Biomed. Res.* 11: 393-404.

Jovin, T. M. (1979) Fluorescence polarization and energy transfer: theory and applications. In M. R. Melamed, P. F. Mullaney and M. L. Mendelsohn (eds.), *Flow Cytometry and Sorting*, pp. 137-65. New York: Wiley.

Kachel, V., Glossner, E. and Schneider, H. (1982) A new flow cytometric transducer for fast sample throughput and time resolved kinetic studies of biological cells and other particles. *Cytometry* 3: 202-12.

Karn, J., Watson, J. V., Lowe, A. D., Green, S. M. and Vedeckis, W. (1989) Regulation of cell cycle duration by c-*myc* levels. *Oncogene* 4: 773-87.

Keij, J. F., Griffioen, A. W., The, T. H. and Rijkers, G. T. (1989) INCA: software for Consort 30 analysis of flow cytometric calcium determinations. *Cytometry* 10: 814-17.

Kendall, M. G. (1948) *Rank Correlation Methods*. London: Griffin.

Kenter, A. L. and Watson, J. V. (1987) Cell cycle kinetics model of LPS-stimulated spleen cells correlates switch region rearrangements with S-phase. *J. Immunol. Meth.* 97: 111-17.

Kenter, A. L., Watson, J. V., Azim, T. and Rabbitts, T. H. (1986) Colcemid inhibits growth during early G1 in normal but not in tumorigenic lymphocytes. *Exp. Cell Res.* 167: 241-51.

Kim, M. and Perry, S. (1977) Mathematical methods for determining cell DNA synthesis rate and age distribution utilizing flow micro-fluorometry. *J. Theor. Biol.* 68: 27–42.

Kosower, N. S., Kosower, E. M., Newton, G. L. and Ranney, A. M. (1979) Bimane fluorescent labels: labeling of normal human red blood cells under physiological conditions. *Proc. Nat. Acad. Sci. USA* 75: 3382–86.

Krisch, K. (1971) Carboxylic ester hydrolyases. In P. D. Boyer (ed.), *The Enzymes*, v. 5, pp. 43–69. New York: Academic Press.

Krishan, A. and Ganapathi, R. (1979) Laser flow cytometry and cancer chemotherapy: detection of intracellular anthracyclines by flow cytometry. *J. Histochem. Cytochem.* 27: 1655–56.

Krishan, A. and Ganapathi, R. (1980) Laser flow cytometric studies on intracellular fluorescence of anthracyclines. *Cancer Res.* 40: 3895–3900.

Krishan, A., Sauerteig, A., Gordon, K. and Swinkin, C. (1986) Flow cytometric monitoring of cellular anthracycline accumulation in murine leukaemia cells. *Cancer Res.* 46: 1768–73.

Krishan, A., Sridhar, K. S., Davila, E., Vogel, C. and Sternheim, W. (1987) Patterns of anthracycline retention modulation in human tumour cells. *Cytometry* 8: 306–14.

Kumar, A. A., Freisheim, J. H., Kempton, R. J., Anstead, G. M., Black, A. M. and Judge, L. (1983a) Synthesis and characterization of a fluorescent analogue of methotrexate. *J. Med. Chem.* 26: 111–13.

Kumar, A. A., Kempton, R. J., Anstead, G. M., Proce, E. M. and Freisheim, J. H. (1983b) High-performance liquid chromatography of methotrexate analogues containing terminal lysine or ornithine and their dansyl derivatives. *Anal. Biochem.* 128: 191–95.

Lam, H. Y. P., Talgoy, M. M. and Goldenberg, G. J. (1980) Uptake and decomposition of chlorozotocin in L5178Y lymphoblasts *in vitro. Cancer Res.* 40: 3950–55.

Lansing-Taylor, D., Waggoner, A. S., Murphy, R. F., Lanni, F. and Birge, R. R. (eds.) (1986) *Applications of Fluorescence in the Biomedical Sciences.* New York: Alan R. Liss.

Laplace, P. S. (1812) *Théorie analitique des probabilités,* Part II, Chapter IV, p. 304. Paris: Courcier. Cambridge University Library bench mark Hh.12.74.

Latt, S. A. (1977) Fluorimetric detection of deoxyribonucleic acid synthesis; possibilities for interfacing bromodeoxyuridine dye techniques with flow cytometry. *J. Histochem. Cytochem.* 25: 913–26.

Latt, S. A., George, Y. S. and Gray, J. W. (1977) Flow cytometric analysis of bromodeoxyuridine substituted cell stained with 33258 Hoechst. *J. Histochem. Cytochem.* 25: 927–34.

Legendre, A. M. (1806) *Nouvelles méthodes pour la détermination des orbites des comètes,* p. 72. Paris: Courcier. Cambridge University Library bench mark XVII.40.24.

Lineweaver, H. and Burke, D. J. (1934) The determination of enzyme disassociation constants. *J. Am. Chem. Soc.* 56: 658–66.

MacDonald, P. D. M. (1975) The mathematical theory of exponentially growing cell populations. In A.-J. Valleron (ed.), *Mathematical Models and Cell Kinetics,* p. 15. Ghent: European Press.

Mann, H. B. and Whitney, D. R. (1947) On a test of whether one of two random variables is stochastically larger than the other. *Ann. Math. Statist.* 18: 50–60.

Martin, J. C. and Swartzendruber, D. E. (1980) Time: a new parameter for kinetic measurement in flow cytometry. *Science* 207: 199–200.

Mendelsohn, M. L. (1962) Autoradiographic analysis of cell proliferation in spontaneous breast cancer of C3H mouse III. The growth fraction. *J. Nat. Cancer Inst.* 28: 1015–29.

Michaelis, L. and Menten, M. L. (1913) Die Kinetik der Invertinwirkung. *Biochem. Z. Berlin* 49: 333–69.

Moran, R., Darzynkiewcz, Z., Staiano-Coico, L. and Melamed, M. R. (1985) Detection of 5-bromodeoxyuridine (BrdUrd) incorporation by monoclonal antibodies: role of the DNA denaturation step. *J. Histochem. Cytochem.* 33: 821–27.

Morgan, S. A., Watson, J. V., Twentyman, P. R. and Smith, P. J. (1989) Flow cytometric analysis of Hoechst 33342 uptake as an indicator of multi-drug resistance in human lung cancer. *Brit. J. Cancer* 60: 282–87.

Moroney, M. J. (1990) *Facts from Figures*. London: Penguin Books.

Nelder, J. A. and Mead, R. (1965) A simplex method for function minimization. *Computer J.* 7: 308–13.

Nicolini, C., Belmont, A., Parodi, S., Lessin, S. and Abraham, S. (1979) Mass action and acridine orange: static and flow cytofluorometry. *J. Histochem. Cytochem.* 27: 102–13.

Nooter, K., Oostrum, R., Jonker, R. R., van Dekken, H., Stokdijk, W. and van den Engh, C. J. (1989) Effects of cyclosporin A on daunorubicin accumulation in multi-drug-resistant P388 leukaemia cells measured by real-time flow cytometry. *Cancer Chemother. Pharmacol.* 23: 296–300.

Nooter, K., van den Engh, C. J. and Sonneveld, P. (1983) Quantitative flow cytometric determination of anthracycline content of rat bone marrow cells. *Cancer Res.* 43: 5126–30.

Nusse, M. (1981) Cell cycle kinetics of irradiated synchronous and asynchronous tumour cells with DNA distribution analysis and BrdUrd-Hoechst 33258-technique. *Cytometry* 2: 70–79.

Ormerod, M. G., Payne, A. W. R. and Watson, J. V. (1987) Improved program for the analysis of DNA histograms. *Cytometry* 8: 637–41.

Parker, J. W. (1988) Flow cytometry in the diagnosis of lymphomas. *Cytometry* Suppl. 3: 38–43.

Pearson, T., Galfre, G., Ziegler, A. and Milstein, C. (1977) A myeloma hybrid producing antibody specific for an allotypic determinant on "IgD-like" molecules of the mouse. *Europ. J. Immunol.* 7: 684–90.

Poisson, S. D. (1837) *Sur la probabilité des jugements*. Paris: Bachelier. Cambridge University Library bench mark Hh.12.53.

Pythagoras (B.C. 600) Τό τετράγωνο τῆς ὑποτεινούσης ἰσοῦται μέ τό ἄθροισμα τῶν τετραγώνων τῶν καθέτων πλευρῶν ἑνός ὀρθογωνίου τριγώνου. Samos.

Quastler, H. and Sherman, F. G. (1959) Cell population kinetics in the intestinal epithelium of the mouse. *Exptl. Cell Res.* 17: 420–38.

Rabbitts, P. H., Watson, J. V., Lamond, A., Fischer, W., Forester, A., Stinton, M. A., Evan, G. I., Atherton, E., Sheppard, R. C. and Rabbitts, T. H. (1985) Metabolism of c-*myc* gene products: c-*myc* mRNA and protein expression in the cell cycle. *Embo. J.* 4: 2009–15.

Rabinovitch, P. S., June, C. H., Grossmann, A. and Ledbetter, J. A. (1986) Heterogeneity among T-cells in intracellular free calcium response after mitogen stimulation with PHA or anti-CD-3. Simultaneous use of indo-1 and immunofluorescence with flow cytometry. *J. Immunol.* 137: 952–61.

Rice, G. C., Bump, E. A., Shrieve, D. C., Lee, W. and Kovacs, M. (1986) Quantitative analysis of cellular glutathione in chinese hamster ovary cells by flow cytometry utilising monochlorobimane: some applications to radiation and drug resistance *in vitro* and *in vivo*. *Cancer Res.* 46: 6105–10.

Rockwell, S. C., Kallman, R. F. and Fajardo, L. F. (1972) Characteristics of a serially transplanted mouse mammary tumour and its tissue culture adapted derivative. *J. Nat. Cancer Inst.* 49: 735–47.

Rosowsky, A., Wright, J. E., Cucchi, C. A., Boeheim, K. and Frei, E. (1986) Transport of a fluorescent antifolate by methotrexate-sensitive and methotrexate-resistant human leukaemic lymphoblasts. *Biochem. Pharmacol.* 35: 356–60.

Rosowsky, A., Wright, J. E., Shapiro, H., Beardsley, P. and Lazarus, H. (1982) A new fluorescent dihydrofolate reductase probe for studies of methotrexate resistance. *J. Biol. Chem.* 257: 14162–67.

Ross, D. D., Joneckis, C. C. and Schiffer, C. A. (1986) Effects of verapamil on the *in vitro* intracellular accumulation and retention of daunorubicin in blast cells from patients with acute non lymphatic leukaemia. *Blood* 68: 83–88.

Ross, D. D., Ordonez, J. V., Joneckis, C. C., Testa, J. R. and Thompson, B. W. (1988) Isolation of highly multidrug resistant P388 cells from drug sensitive P388/S cells by flow cytometric cell sorting. *Cytometry* 9: 359–67.

Rotman, B. and Papermaster, B. W. (1966) Membrane properties of living mammalian cells as studied by enzymatic hydrolysis of fluorogenic esters. *Proc. Nat. Acad. Sci. USA* 55: 134–41.

Schimke, R. T. (1984) Gene amplification, drug resistance and cancer. *Cancer Res.* 44: 1735–42.

Schuette, W. H., Shackney, S. E., MacCollum, M. A. and Smith, C. A. (1983) High resolution method for the analysis of DNA histograms that is suitable for the detection of multiple aneuploid G1 peaks in clinical samples. *Cytometry* 3: 376–86.

Siegel, S. and Castellan, N. J. (1988) *Non-parametric Statistics for the Behavioural Sciences,* 2nd ed., pp. 144–51. New York: McGraw-Hill.

Sirontak, F. M., Moccio, D. M., Kelleher, L. E. and Goutas, L. J. (1981) Relative frequency and kinetic properties of transport-defective phenotypes among methotrexate-resistant L1210 clonal cell lines derived *in vitro. Cancer Res.* 41: 4447–52.

Sonneveld, P. and van den Engh, C. J. (1981) Differences in uptake of adriamycin and daunomycin by normal BM and acute leukaemia cells determined by flow cytometry. *Leukaemia Res.* 5: 251–57.

Southern, E. M. (1975) Detection of specific sequences among DNA fragments by gel electrophoresis. *J. Mol. Biol.* 98: 503–17.

Steel, G. G. (1968) Cell loss from experimental tumours. *Cell and Tissue Kinet.* 1: 193–207.

Steel, G. G. (1973) The measurement of the intermitotic period. In M. Balls and F. S. Billett (eds.), *The Cell Cycle in Development and Differentiation.* Cambridge: Cambridge University Press.

Steel, G. G., Adams, K. and Barrett, J. C. (1966) Analysis of cell population kinetics of transplanted tumours of widely differing growth rate. *Brit. J. Cancer* 20: 784–89.

Steinkamp, J. A. (1983) A differential amplifier circuit for reducing noise in axial light loss measurements. *Cytometry* 4: 83–87.

Stryer, L. (1978) Fluorescence energy transfer as a spectroscopic ruler. *Ann. Rev. Biochem.* 47: 819–46.

"Student" (1908) The probable error of a mean. *Biometrica* 6: 1–25.

Studzinski, G. P., Brulei, Z. S., Feldman, S. C. and Watt, R. A. (1986) Participation of c-*myc* protein in DNA synthesis of human cells. *Science* 234: 467–70.

Takahashi, M. (1966) Theoretical basis for cell cycle analysis, I. Labelled mitosis wave method. *J. Theoret. Biol.* 13: 202–11.

Takahashi, M. (1968) Theoretical basis for cell cycle analysis, II. Further studies on labelled mitosis method. *J. Theoret. Biol.* 18: 195–209.

Takahashi, M., Hogg, J. and Mendelsohn, M. L. (1971) The automated analysis of FLM curves. *Cell and Tissue Kinet.* 4: 505–18.

Taylor, I. W. (1980) A rapid single step staining technique for DNA analysis by flow microfluorimetry. *J. Histochem. Cytochem.* 28: 1021–24.

Teresima, T. and Tolmach, L. J. (1963) Growth and nucleic acid synthesis in synchronously dividing populations of HeLa cells. *Exptl. Cell Res.* 30: 344–62.

Traganos, F., Darzynkiewicz, Z., Sharpless, T. and Melamed, M. R. (1977) Simultaneous staining of ribonucleic and deoxyribonucleic acid in unfixed cells using acridine orange in a flow cytofluorimetric system. *J. Histochem. and Cytochem.* 25: 46–56.

Traganos, F., Darzynkiewicz, Z., Sharpless, T. and Melamed, M. R. (1979) Erythroid differentiation of Friend leukaemia cells as studied by acridine orange staining and flow cytometry. *J. Histochem. Cytochem.* 27: 382–89.

Tsuro, T., Iida, H., Tsuragoshi, S. and Sakurai, Y. (1982) Increased accumulation of vincristine and adriamycin in drug-resistant P388 tumour cells following incubation with calcium antagonists and calmodulin inhibitors. *Cancer Res.* 42: 4730–33.

van Ewijk, W., van Soest, P. L., Verkerk, A. and Jonkind, J. F. (1984) Loss of antibody binding to prefixed cells: fixation parameters for immunocytochemistry. *Histochem. J.* 16: 179–93.

Visser, J. W. M., van den Engh, G. J. and van Bekkum, D. M. (1980) Light scattering properties of murine haemopoetic stem cells. *Blood Cells* 6: 391–407.

Watson, J. V. (1977a) Fluorescence calibration in flow cytofluorimetry. *Brit. J. Cancer* 36: 396.

Watson, J. V. (1977b) The application of age distribution theory in the analysis of cytofluorimetric DNA histogram data. *Cell and Tissue Kinet.* 10: 157–69.

Watson, J. V. (1978) A linear transform of the multi-target survival curve. *Brit. J. Radiol.* 51: 534–38.

Watson, J. V. (1980) Enzyme kinetic studies in cell populations using fluorogenic substrates and flow cytometric techniques. *Cytometry* 1: 143–51.

Watson, J. V. (1991) *Introduction to Flow Cytometry,* Chapter 15. Cambridge: Cambridge University Press.

Watson, J. V. (in press) Proof without prejudice revisited; ratio analysis of means plus Kolmogorov-Smirnov statistics for immunofluorescence histogram analysis. *Cytometry.*

Watson, J. V., Chambers, S. H. and Smith, P. J. (1987) A pragmatic approach to the analysis of DNA histograms with a definable G1 peak. *Cytometry* 8: 1–8.

Watson, J. V., Chambers, S. H., Workman, P. and Horsnell, T. S. (1977) A flow cytometric method for measuring enzyme reaction kinetics in populations of single cells. *Febs Lett.* 81: 179–92.

Watson, J. V., Cox, H., Hellon, C., Workman, P. and Dive, C. (1988) Time-resolved flow cytometric measurement of dynamic events in the interval 1 to 20 seconds in whole cells. *Cytometry* Suppl. 2: 93a.

Watson, J. V., Cox, H., Hellon, C., Workman, P. and Dive, C. (in press) Stop-flow cytometry in flow: measurement of "fast" dynamic events in the interval 1 to 20 seconds in whole cells. *Cytometry.*

Watson, J. V., Morgan, S. A. and Dive, C. D. (1989) Drug resistance analysis in populations of individual cells using flow cytometry. In F. Sharp, W. P. Mason and R. E. Leake (eds.), *Ovarian Cancer; Biological and Therapeutic Challenges,* Chapter 14. London: Chapman and Hall.

Watson, J. V., Morgan, S. A. and Smith, P. J. (1989) Analysis of the Hoechst-33342 fluorescence emission spectrum in individual cells with reference to multi-drug resistance. *Protides of the Biological Fluids* 36: 359–72.

Watson, J. V., Sikora, E. K. and Evan, G. I. (1985) A simultaneous flow cytometric assay for c-*myc* oncoprotein and cellular DNA in nuclei from paraffin embedded material. *J. Immunol. Meth.* 83: 179–92.

Watson, J. V., Stewart, J., Evan, G., Ritson, A. and Sikora, K. (1986) The clinical significance of flow cytometric c-*myc* oncoprotein quantitation in testicular cancer. *Brit. J. Cancer* 53: 331–37.

Watson, J. V. and Taylor, I. W. (1978) Cell cycle analysis *in vitro* using flow cytofluorimetry after synchronization. *Brit. J. Cancer* 36: 281–87.

Watson, J. V. and Walport, M. J. (1985) How does flow cytometry interpret Gaussian distributed information. *J. Immunol. Meth.* 77: 321–30.

Watson, J. V., Workman, P. and Chambers, S. H. (1978) Differences in esterase activity between exponential and plateau growth phases of EMT6/M/CC cells monitored by flow cytofluorimetry. *Brit. J. Cancer* 37: 397–402.

Watson, J. V., Workman, P. and Chambers, S. H. (1979) An assay for plasma membrane phosphatase activity in populations of individual cells. *Biochem. Pharmacol.* 28: 821–27.

White, R. A. and Meistrich, M. L. (1986) A comment on: A method to measure the duration of DNA synthesis and the potential doubling time from a single sample. *Cytometry* 7: 486–90.

Whittaker, E. and Robinson, G. (1960) In *The Calculus of Observations,* 4th ed., p. 206. London: Blackie.

Wilcoxon, F. (1945) Individual comparisons by ranking methods. *Biometrics Bull.* 1: 80–87.

Wilson, G. D., McNally, N. J., Dunphy, E., Karcher, H. and Pfragner, R. (1985) The labelling index of human and mouse tumours assessed by bromodeoxyuridine staining *in vitro* and *in vivo* staining and flow cytometry. *Cytometry* 6: 641–47.

Workman, P., Cox, H. and Watson, J. V. (1988) Glutathione metabolism in human lymphocytes by flow cytometry. *Cytometry* Suppl. 2: 47a.

Workman, P. and Watson, J. V. (1987) Flow cytoenzymology of gutathione metabolism. *Cytometry* Suppl. 1: 48a.

Wright, J. E., Rosowsky, A., Boeheim, K., Cucchi, C. A. and Frei, E. (1987) Flow cytometric studies of methotrexate resistance in human squamous carcinoma cell cultures. *Biochem. Pharmacol.* 36: 1561–64.

Yates, F. (1934) Contingency tables involving small numbers and the χ^2 test. *J. Royal Statistical Soc.* Suppl. 1: 217–35.

Young, I. T. (1977) Proof without prejudice: use of the Kolmogarov-Smirnov test for the analysis of histograms from flow systems and other sources. *J. Histochem. Cytochem.* 25: 935–41.

Zeitz, S. and Nicolini, C. (1978) Flow microfluorometry and cell kinetics: a review. In A.-J. Valleron and P. D. M. MacDonald (eds.), *Biomathematics and Cell Kinetics,* p. 357. Amsterdam: Elsevier.

Index